Quick Tricks
for Science

by Barbara F. Backer
illustrated by Marilynn G. Barr

Dedicated with love to my mother, Ruth J. Feldman,
who shares with me her love for nature.

Publisher: Roberta Suid
Design & Production: Standing Watch Productions
Cover Design: David Hale

CONTENTS

INTRODUCTION

When you ask adults to think of a happy memory from childhood, most will give an answer that relates to the outdoors. Children's first look at science often comes while playing outdoors, observing nature and playing with plants. This book focuses on learning science processes by observing plants, both indoors and out.

To help children learn about plants, this book contains many "Quick Tricks"—simple activities with everyday items. Wherever children live, they can observe differences in plants in each season of the year. Observing these changes and also noticing the things that don't change are all part of scientific investigation.

The Quick Tricks will help your children learn that most plants have three main parts—the root, the stem, and the leaves—and that many plants also have flowers that turn into fruits with seeds. They'll learn that some plant parts and seeds are edible and that it's dangerous to eat others. Help them understand that they can only eat plants and seeds that a grown-up identifies as safe.

Preschool children learn best from hands-on experiences with learning materials. Therefore, your children will be encouraged to handle plant materials and to talk about what they learn.

Be flexible when working with children; be ready for the unexpected. If you go outdoors to gather leaves and the children discover blossoming wildflowers, take advantage of the learning opportunity that presents itself. Inspect the flowers; discuss their colors and shapes.

The first chapter, "General Materials," tells how to make or use items that will be used in various "tricks" throughout the book. Read and use this chapter first. After that, you can use the Quick Tricks in any order.

These Quick Trick activities are meant to be enjoyed in a casual, informal setting and atmosphere. The emphasis is on having a positive learning experience, not in getting a right or wrong answer. If children offer incorrect information, ask them how they figured that out. Suggest other ways to explore, and ask questions that lead the children to accurate information.

The final chapter includes Quick Tricks that help you spread the plants theme across the curriculum. These activities help children increase their math and language skills and enhance their visual discrimination.

Most chapters include a letter for parents. The letters inform parents of what the children are learning at school and offer a Quick Trick for continuing the learning at home. To keep the language simple, chapters alternate between referring to children as "he" or "she."

Quick Tricks for Science ©2001 Monday Morning Books, Inc.

GENERAL MATERIALS

The Quick Tricks in this chapter focus on materials that you will use in future chapters.

Wow! That's Big!
A Quick Trick with a Magnifying Glass

Gather These Materials:
magnifying glasses of various sizes and strengths

Where: anywhere

How: Introduce children to the magnifying glasses. Explain that they come in different sizes and shapes, but all of them do the same thing. They help us see things more closely by making the items appear larger.

You may need to work one-to-one to help children understand how to use the glasses. Children who have seen cartoon pictures of investigators want to hold the glass up to their eye to look into it. Help them focus on an object then bring the glass close to the object in a line between their eye and the object. Show them how to move the glass back and forth slowly toward the object until they "see" an enlarged image. Let them practice with various magnifying lenses.

<u>Caution</u>: Be careful using magnifying glasses outdoors. If the lens is held in one position with the sun coming through the lens, it concentrates the sun's light and heat onto one place. That place can quickly catch fire.

Variation:
Have children look at a leaf without using a magnifying glass and have them draw the leaf. Now, have them examine the leaf while using a magnifying glass. Have them draw the leaf as it appeared with the magnifying glass. Discuss how the pictures are different. Provide the children many opportunities to explore with magnifying glasses.

This activity also helps children learn about:
tools that can help us when we explore science.

Making Clipboards
A Quick Trick with Cardboard and Paper Clips

Gather These Materials:
yardstick
heavy cardboard
sharp scissors or a craft knife, for adult use only
spring-type (Bulldog) paper clips, one for each child

Where: anywhere

How: <u>This is an adult activity.</u>
    For each child, measure the cardboard and cut a 9-inch by 12-inch (22.5-cm by 30-cm) piece for each child. Clip a large, spring-type paper clip to one narrow end of the cardboard. (Substitute two jumbo, vinyl-clad paper clips if necessary.)
    To use the clipboards, use the clips to hold plain paper in place on the boards.

Variation:
Use a magazine instead of cardboard. Clip the spring-type clip to the magazine to hold plain paper on the clipboard.

Quick Tricks for Science ©2001 Monday Morning Books, Inc.

Junior Journal
A Quick Trick with Paper Bag

Gather These Materials:
adult scissors, for adult use only
paper bag for each child
plain paper, 12 pieces for each child
spring-type clothespins or jumbo paper clips
hole punch
yarn
twist ties
markers

Where: at a table

How: The adult cuts open and then flattens paper bags into one large sheet. From this, cut pages the size of a sheet of plain paper. Cut two pieces for each child.

Give each child 12 sheets of paper and two pieces of paper-bag paper. Demonstrate how to stack the papers with one piece of paper-bag paper on the top and one on the bottom. Have each child do the following: With a partner, hold the sheets of paper steady while the partner clamps the papers together by putting a clothespin on each of the four edges of the stack. Repeat for the partner's stack of papers.

The adult does the following with each child: Punch holes about 1 inch (2.5 cm) apart along one long edge of each paper stack. Tie one end of an 18-inch (45-cm) length of yarn to the center of a twist tie. Fold the twist tie in half, and then twist it tightly shut (for a sewing needle). Tie the other end of the yarn through an end hole in the paper stack.

Have each child sew yarn through the holes of his paper stack, sewing a "binding" for the book. Don't worry about precision or neatness. Just remind the children to always sew through the "next" hole and to pull the yarn snugly between each stitch. Help the children cut off the "needle" and tie off the remaining end of the yarn. Have the children copy the word "Journal" onto the covers of their books.

Variation:
Stack a few index cards and staple one edge to make a small, hand-size journal.

This activity also helps children learn about:
a creative way to reuse a paper bag.

Making a Field Guide
A Quick Trick with Construction Paper and Plain Paper

Gather These Materials:
plain paper, 6 pieces for each child
construction paper slightly larger than the plain paper
stapler
markers

Where: anywhere

How: Give each child six pieces of plain paper and one piece of construction paper. Show them how to fold each sheet of plain paper in half and then open up all six sheets. Now place the six sheets in a stack. Fold the construction paper in half and then open it and insert the stack of plain paper. Align the papers, then close the entire stack into a book. Help the children use the stapler to staple the book in several places near the folded edge. They now have a 12-page blank book with a construction paper cover. Have them use markers to decorate the front of their field guides.

Variation:
Obtain a variety of field guides from the library. Include guides to flowers, plants, and trees. Show them to the children and explain how they are used. (When you find a natural item, you can look for its picture in the field guide and learn its name.) Explain that the children will use their blank books to create a personal field guide. Suggest that they decorate the book's front with pictures of plant items (similar to the library field guides).

The children will use their field guides to draw pictures of items they've seen (trees, leaves, flowers) and label them with your help. Later they can refer to their guides to remind them of natural items' names.

This activity also helps children learn about:
using a field guide as a tool for exploring nature.

A PLANT'S NEEDS

Plants need light, water, food, and air to survive. The Quick Tricks in this chapter explore these needs and ask the question, "Can you have too much of a good thing?"

Do Plants Need Water?

A Quick Trick with a Paper Cup

Gather These Materials:
2 similar potted plants
a small paper cup

Where: indoors

How: With the children, explore the two plants. (These can be bean plants growing from seeds the children have planted.) Determine how they are similar: size, type of plant, health.

Ask the children if plants must have water to live and grow. After discussion, ask, "How can we find out?" With questions, lead them to suggest watering one plant and not watering the other. Have them predict what will happen to each plant. Record their predictions.

With thumb and forefinger, pinch the rim of the cup, forming a small pouring spout. For the next few days/weeks, let children use this "watering cup" to water just one of the plants. Be certain that the plants are potted with good drainage to prevent "drowning" the plant that receives water.

Every few days, observe and compare the plants. Are there any changes? (The watered plant should be growing and the unwatered plant should be drying up.) Compare the changes with the children's predictions. Continue the experiment until the unwatered plant dies. Compare the plants and then discuss and record your observations.

Variation:
Use more than two plants. Water several and don't water the others. When the unwatered plants begin to fade, ask children what might happen if you start to water one of them. Label that plant. Record their predictions and begin watering that plant also. In a few more days, label and begin watering another unwatered plant. Continue observing the plants and discuss your results.

This activity also helps children learn about:
how plants change when they don't receive enough water.

Do Plants Need Light?
A Quick Trick with a Dark Cabinet or Closet

Gather These Materials:
2 similar potted plants
a dark cabinet or closet
crayons/markers
student journals

Where: indoors

How: With the children, explore the two plants as in the previous activity. Determine the many ways they are similar.

Ask the children if plants need light to live and grow. After discussion, ask, "How can we find out?" With questions, lead them to suggest leaving one plant in or near a window and putting the other in a dark place (a cabinet or closet). Have them predict what will happen to each plant. Record their predictions.

Water both plants on the same schedule and compare them every few days. Discuss any changes that occur. What happens to the plant that has light and to the plant that has no light? Refer to the children's predictions. Did the outcome match the predictions? Have the children draw pictures of the two plants in their journals. Write their dictated descriptions of the plants beside the drawings.

Variation:
Experiment to see how similar plants react to different amounts of light. Gather some similar potted plants. Encourage the children to find differently lit locations where they can put the plants. Suggest putting one in full, hot sun; one in full shade, one in bright, but indirect light. Have the children observe and compare the plants over time. Discuss the observations.

This activity also helps children learn about:
predicting results and comparing the predictions with the actual results.

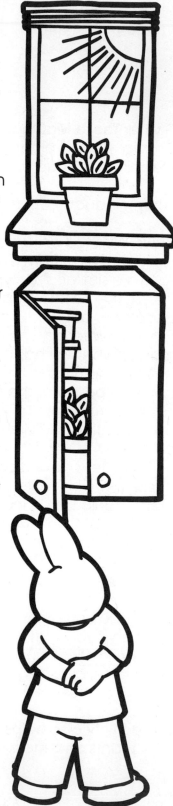

Quick Tricks for Science ©2001 Monday Morning Books, Inc.

Do Plants Seek Light?
A Quick Trick with a Window

Gather These Materials:
healthy potted plant(s)
permanent marker
masking tape
markers
journal books

Where: indoors

How: Examine and discuss the potted plant. Observe its growth. Remind the children that a plant needs light to grow. Tell the children that you are going to place the plant near a sunny window and that you will always keep one side of the plant toward the window and the other side toward the room. Ask the children to predict if the plant will stay healthy. Do they think it will change in any way?

Draw a sun on a piece of masking tape and stick it to the side of the pot that will face the sunny window. Draw a dark circle on a piece of masking tape and stick it to the opposite side of the pot. Have the children draw pictures in their journals of the plant and write the date on their pictures.

Place the pot in the window. Observe it every few days. What is happening? Compare the plant to the children's pictures. As the children notice the plant growing toward the sunny window, have them draw the plant again and date their pictures. Continue this for a few weeks. Discuss and record all changes.

Variation:
Have the children predict what might happen if you now turn the pot around so the "sun" label faces the inside of the room and the "dark" label faces the window. Record their predictions, then conduct the experiment. Observe, discuss, and record the results.

This activity also helps children learn about:
changes over time.

Adopt a Plant
A Quick Trick with Potted Plants

Gather These Materials:
potted plants, indoor variety
plant food
water
premade blank books
crayons, markers, pens
camera and copy machine (optional)

Where: indoors

How: Show the children several potted, indoor plants. Compare the plants and discuss the similarities and differences. Have the children choose their favorite plant. Divide children into small groups according to their favorite. Each group then "adopts" its plant.

Children in each group take turns caring for "their" plant, feeding it, watering it, and making sure it gets proper light. Explain that each child will keep a journal on his adopted plant. Have them draw a picture of their plant the first day. They may also want to draw a close-up of the plant's leaves or flowers. Have the children measure the plant every two weeks and record its growth. Have them draw pictures of new growth. If desired, take photos of the plants and make color photocopies of these for the children to use on their books' covers.

Variation:
After the children learn about a plant's needs, show them sick plants. Discuss ways they might help the plants get healthy. Divide into groups, as above. Have each group adopt their favorite plant and try to revive it with proper light, proper watering, and plant food. If repotting or dividing the plant is necessary, let the children help cut off dead growth and sick roots. Have the children observe the plant regularly and have them keep a journal that records the plant's "treatments" and the results.

This activity also helps children learn about:
observing for changes over time and recording them.

I'm Not <u>That</u> Thirsty
A Quick Trick with a Measuring Cup

Gather These Materials:
healthy potted plant
measuring cup
water

Where: anywhere

How: Have the children examine a potted plant to see if they agree that it is healthy. Discuss a plant's needs, including the need for water. Ask children what will happen if you water a plant a lot every day. Record their predictions on a chart.

Have a different child give the plant 1/2 cup (118 ml) of water each day. Make a mark on the chart for each watering. Encourage the children to observe the plant. Record their observations on another chart. Is water collecting in the pot? Is it spilling over the top? How does the soil feel? What is happening to the plant? Compare this information to the predictions on the first chart.

When the plant begins to die, discuss what may be killing it. Continue watering the plant daily, and examine it after it dies. Remove it from the soil and examine the roots. Compare all information to the predictions on the chart.

Variation:
If the children become alarmed about the plant's health, let them vote whether or not to end the experiment. What do they think might restore the plant to health? Vote on what to do, and observe the plant for changes.

This activity also helps children learn about:
making and checking predictions.

Dear Parents,

At school we are learning about a plant's needs. We've been experimenting with many plants, and we've learned that plants need light and water to grow—but not too much water!

You can extend this learning at home by doing the "Quick Trick" below:

Quick Trick with a Potato

Gather These Materials:
sharp scissors, for adult use only
empty 3-quart (3-liter) plastic bottle
sharp implement, for adult use only
used large plastic jug
knife
potato with eyes
5-inch (12.5-cm) square cut from discarded
pantyhose or knee-highs
spoon
soil

Where: at a covered surface or outdoors

How: The adult does the following: Cut the 3-quart (3-liter) bottle 5 inches (12.5 cm) from the bottom. Discard the top section. Use a sharp implement to punch five small holes in the bottom of the 5-inch (12.5 cm) section. Cut the plastic jug 1 inch (2.5 cm) from the bottom and discard the top section. Cut a 2-inch (5-cm) piece from the potato. Be sure it contains one or two eyes.

The child does this: Flatten the hosiery in the bottom of the 3-quart (3-liter) container's bottom. (This keeps the soil from washing through the holes.) Spoon about 2 inches (5 cm) of soil on top of the hosiery. Place the potato piece in the container with the cut side down and cover the piece with soil. Gently tap the soil in place. Place the container in the jug's bottom (to catch run-off water). Water the plant.

With proper care, the potato will sprout and send out long vines. When the plant has a few inches of growth, help your child transplant it into a garden, if possible. Over the following year, it may produce new potatoes under the ground.

LEAVES

All plants have leaves, which come in many sizes and shapes. In these Quick Tricks, your children will expand their vocabularies while they explore and describe leaves from many kinds of plants. They may be surprised to learn that they often eat leaves. They'll also learn how to use leaves in creative ways.

Looking at Leaves
A Quick Trick with a Magnifying Glass

Gather These Materials:
a small bag for each child
magnifying glasses
markers/crayons
nature journals

Where: outdoors to gather leaves
anywhere for exploring the leaves

How: Go on a leaf hunt. (Before you go, be sure the area has no plants that are poisonous to the touch.) Encourage the children to gather leaves they find on the ground and place them in their bags. Remind them to find a variety of sizes, shapes, and colors. The adult gathers a few live leaves from trees, bushes, and other plants, as well.

In a group, have the children examine their leaves and discuss how the leaves are similar and different. Use words like *broad, long, sturdy, delicate, smooth, rough, striped,* and *wrinkled.* Give the children magnifying glasses and continue the exploration and discussion.

Have the children make a picture of one or more of their leaves to put in their nature journals. Write the children's descriptive words on their pictures.

Variation:
Have the children select five leaves from their collections and place them in order from largest to smallest.

This activity also helps children learn about:
recording observations.

Leafy Lines
A Quick Trick with Crayons

Gather These Materials:
leaves from various plants
magnifying glass
newspaper
plain paper
masking tape
crayons with paper removed
scissors
student journals

Where: outdoors to gather leaves
at a table for the rubbings

How: Have your children gather leaves from a variety of plants,
including trees, shrubs, and flowers. Examine the leaves close-up.
Find the heavy lines in each leaf, the leaf's veins. Use a magnifying
glass and look at the smaller lines all through the leaf. Discuss how
these are connected to the larger lines. Examine a variety of leaves
to see that all of them, large and small, have veins.

To make a rubbing of a leaf and its veins, have each child do the
following: Help cover the table with newspaper. Place a leaf face-
side-down on the table and put a piece of plain paper over the
leaf. Use masking tape to hold the paper in place on the table.
Briskly rub the side of the crayon over the paper. The leaf's impression will show on the paper, highlighting its veins.

Have the children cut around their leaf rubbings and glue them
into their journals.

Variation:
Use colored chalk or colored pencils instead of the crayons.

This activity also helps children learn about:
developing observation skills.

Edible Leaves I
A Quick Trick with Oregano

Gather These Materials:
live oregano plant
bottle of dried, crushed oregano leaves
tomato sauce
English muffin halves
shredded mozzarella cheese
toaster oven

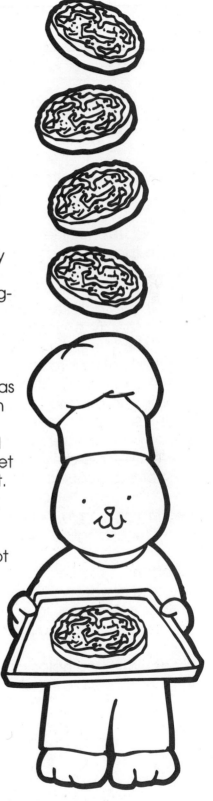

Where: at a table

How: Show your children a living oregano plant. (These are available at plant nurseries.) Let them smell it and gently rub a leaf between their thumb and their index finger. This will leave the oregano aroma on their fingers. Talk about how the plant smells and feels. Ask if the aroma is familiar. Some children may mention spaghetti or pizza.

Tell the children that oregano is a strong-smelling and strong-tasting herb. Cooks use its leaves to flavor foods.

Shake a small amount of the bottled oregano leaves into each child's hand. As the children explore them, explain that they are dried, crushed oregano leaves.

A few at a time, have the children make single-serving pizzas by spreading tomato sauce on an English muffin half. Let them sprinkle on a small amount of crushed oregano and top the muffin half with shredded mozzarella cheese. Toast or top broil the muffins in a toaster oven until the cheese begins to melt. Let the cheese cool before giving the pizzas to the children to eat.

Don't force anyone to make, taste, or eat the pizza. Some children will enjoy making it, but won't want to eat it.

Caution: Remind children that they should not eat leaves they find on plants unless an adult says the leaves are safe. Not all leaves are edible.

Variation:
Smell and taste other growing herbs, then taste a food that uses those herbs in preparation. Some suggestions: dill and dill pickles; mint leaves and mint candies.

This activity also helps children learn about:
combining foods and seasonings to make a different food.

Edible Leaves 2
A Quick Trick with Salad Greens

Gather These Materials:
a variety of leafy salad greens, whole heads if possible
basin and water
large, clean towels
large bowl
cooking spoons
small bowls, 1 for each child
mild salad dressings
forks

Where: at a table to wash greens and prepare and eat salad

How: Show the children a variety of leafy salad greens: romaine, escarole, red-leaf lettuce, iceberg lettuce, spinach, and others. Explain that these are all plant leaves. Show them the bottoms of the greens, where the leaves were cut away from the bottom of the plant.

Help the children separate and wash the greens and then spread them on large towels. Show them how to roll up the towels and gently blot the moisture from the leaves. Unroll the towels and tear the leaves into small, bite-size pieces. Use a lot of the mild greens—iceberg lettuce and red-leaf lettuce—and only a little of the romaine and other "sharp" greens. Have the children put the torn greens into a large bowl. After they stir to mix the greens, let them serve themselves some of the resulting salad.

Provide a selection of mild dressings to choose from and forks for eating. Enjoy the salad. Don't force anyone to taste or eat the food. All of the children are learning from watching and taking part in the preparation. Eating and tasting are not required.

Variation:
Offer the children a variety of washed greens and single-size servings of dressing in small paper cups. Let the children sample the greens, dipping them first in the dressing if they desire.

This activity also helps children learn about:
preparing foods.

Edible Leaves 3
A Quick Trick with Spinach

Gather These Materials:
raw spinach leaves
towel
cooked spinach (canned, frozen, or fresh)
plates/small cups for serving
forks

Where: anywhere

How: Wash raw spinach leaves and pat dry. Show the children the raw leaves and see if they can guess what the leaves might be. (Don't tell them the answer.) Let them examine the leaves, and guide the discussion so the children discuss that these are edible plant leaves. Have them smell, feel, and (if they desire) taste the leaves, and have them save some of the leaves for later.

Next, tell the children you have some of the same kind of leaves, but they have been cooked. Give each child a taste-size portion of the cooked spinach. Let the children smell, touch, and then taste the cooked leaves. Don't force anyone to taste or eat the food. Children who don't want to taste it may agree to smell it and touch it. Can the children identify the food? Discuss the fact we eat both raw and cooked leaves from the spinach plant.

Have the children look at the raw leaves they've saved. How are these the same as or different from the cooked leaves?

Variation:
Repeat the activity with collard greens, mustard greens, cabbage, beet greens, or other leafy, green vegetables. Examine and taste them both raw and cooked.

This activity also helps children learn about:
how heat changes things during cooking.

Leaf People
A Quick Trick with Leaves

Gather These Materials:
leaves of many sizes and shapes
variety of plant materials: twigs, small flowers and weeds, grasses,
acorns, nuts, small pinecones

Where: anywhere

How: Use a broad leaf as a base for creating a leaf person. With a
thin twig, poke small holes in the leaf. Stick flowers, acorns, tiny
leaves, and other natural items into the holes to make features:
eyes, nose, ears, and so on. Use small leaves or small twigs for arms
and legs. Make leaf people of all sizes to create a leaf family.

Variations:
Make leafy animals for your leaf people.
Using leaves, paper, and glue, make a collage of leaf people.

This activity also helps children learn about:
creative thinking.

Quick Tricks for Science ©2001 Monday Morning Books, Inc.

Leafy Impressions
A Quick Trick with Leaves and Modeling Dough

Gather These Materials:
leaves
modeling dough
Vaseline
old (inkless) ballpoint pen

Where: outdoors to gather leaves
at a table for the activity

How: Have each of the children gently gather a few leaves from living trees, bushes, or other plants. Remind them that the plants need their leaves, so they should gather no more than one leaf from each specimen. At a table, have the children examine their leaves and select the leaf with the most prominent veins. Put away the unselected leaves, and have the children set down their selected leaves.

Give each child a fist-size piece of modeling dough. Have the children manipulate the dough until it is soft and pliable. Now have them flatten the dough into a "pancake."

Show the children how to cover the undersides of their leaves with Vaseline and how to lay the Vaseline-covered leaf on the dough pancake, bottom down. Have them press the leaves gently into the pancakes. What do they find when they remove the leaves from the dough? Use the ballpoint pen to "write" each leaf's name in its pancake. Display these "fossilized leaf records" together in an area labeled "Leaf Museum." Encourage the children to examine and compare these leafy "fossils."

Variation:
Use this same technique to make molds of seeds your children find: acorns, maple tree "helicopters," and sunflower, marigold, apple, and green pepper seeds, for example.

This activity also helps children learn about:
how fossils were formed.

Dear Parents,

In class we are observing leaves, noticing their colors, sizes, and shapes. We have smelled the aroma of fresh and dried herbs and have tasted cooked herbs on other foods. We've learned that lettuce, cabbage, and spinach are leaves we can eat.

You and your child can share the pleasure of exploring leaves you see outdoors and in. Your child will enjoy making permanent foil prints of some of the leaves you find. Gather leaves in a variety of shapes and sizes, then try this "Quick Trick":

A Quick Trick with Aluminum Foil

Gather These Materials:
a variety of thick leaves from trees, bushes, and other plants (avoid delicate, lacy leaves)
aluminum foil
glue (optional)
construction paper (optional)
picture frame (optional)

Where: outdoors to gather the leaves
at a table to make the impressions

How: Have your child do the following: Place a leaf face down on the table. Cover it with a piece of aluminum foil that is larger than the leaf. With the fingers and thumb, gently press the foil down and around the leaf. Smooth from the center of the leaf to the outside. Press down firmly around the leaf's edges. Remove the leaf and trim the foil, leaving at least a 1-inch (2.5-cm) border around the leaf. If desired, glue the foil impression to a piece of construction paper that is larger than the foil. Finished pages can be gathered into a book or framed.
　　To frame the leaf impressions, cut the construction paper to fit the frame and cut the impressed foil so you have a 1-inch (2.5-cm) border of construction paper.

STEMS

> The stem might be the most ignored part of most plants. We tend to focus on the leaves and flowers. But the stem holds up the leaves and carries nutrients throughout the plant. These Quick Tricks help children explore stems.

Traveling Water
A Quick Trick with Celery

Gather These Materials:
clear, plastic jar
water
red or blue vegetable dye (food coloring)
bunch of fresh, crisp celery with leaves attached

Where: at a table

How: Show the children the celery and where the celery bunch was cut from the rest of the plant. Give each child a stalk of celery to examine. Identify the leaves and the stem of the celery plant.

Explain that plants need water to survive. Ask the children if they know how water travels in a plant. Explain that a growing plant has roots that take water from the soil; the water then travels through the stem and into the leaves and flower blossoms. Ask if they'd like to see how the water travels.

Partially fill the jar with water. Add enough vegetable dye to give the water a dark color. Snap the bottoms off of several stalks of celery. Immediately place the celery into the colored water. In a few hours, the children will see the dark water moving up the celery stalk (stem). Snap the bottom inch (2.5 cm) from one stalk and show them the dark-colored places in the cross-section. These parts of the stem transport water.

Continue to observe the celery, adding more colored water if necessary. You may be able to see the dark-colored water in the leaves' veins.

Variation:
After finishing this activity, slit a celery stalk in half, lengthwise, stopping about 4 inches (10 cm) from the top of the stalk. Place two clear, plastic tumblers side by side. Put water in each. Color the water in one with vegetable dye so it is dark red, and color the other water dark blue. Carefully place the celery stalk over the tumblers' rims, putting one side in the red water and one in the blue water. Ask the children what will happen. Observe the celery for several days. Compare the results with the predictions.

This activity also helps children learn about:
the fact that the vegetables we eat contain a lot of water.

Stems and More Stems
A Quick Trick with Words

Gather These Materials:
a variety of flowers with their stems attached
a chart
marker

Where: anywhere

How: In a group, have the children explore and compare the flowers' stems. Encourage them to describe the stems. Record all of their descriptive words on a chart. Ask questions like these to expand their descriptions: "How is this stem the same as that one?" "How are these stems different?" "What colors do you see?" "What can you tell me about size?"

Variation:
Make a word card for each word the children suggested. Let the children match the word cards to the words on the chart.

This activity also helps children learn about:
observing for similarities and differences.

Quick Tricks for Science ©2001 Monday Morning Books, Inc.

The Surprise in the Stem
A Quick Trick with a Branch

Gather These Materials:
a branch from a shrub or a small tree—gather this at winter's end
when the branch is bare
clippers
magnifying glass (optional)
container to hold the branch
water

Where: outdoors to gather the branch
indoors for the branch to sprout

How: At winter's end, cut a bare branch from a shrub or from a small tree. Examine the branch. Do the children think it is alive?
Place the branch in a container of water. Bring the branch and water inside. Replenish the water regularly. In time the branch will sprout buds that will open to reveal leaves. Several branches in an attractive container make a pretty bouquet or natural arrangement.

Variation:
Have the children draw pictures of the branch when you first bring it indoors, when the buds begin to grow, and when the buds put out young leaves.

This activity also helps children learn about:
observing changes over time.

Painting with Twigs
A Quick Trick with Paints and Twigs

Gather These Materials:
tempera paints
flat containers for paint
twigs in a variety of sizes
paper
black marker

Where: anywhere

How: Put tempera paint in flat containers. (Disposable aluminum roasting pans or Styrofoam grocery trays work well.) Have the children gather a variety of twigs. Show them how to dip an entire twig in paint and lay it sideways on the paper. Have a child pick up one edge of the paper so that the twig rolls on the paper, leaving a paint design. Repeat using a different twig and a different color of paint. Continue until the artist has a pleasing design. Have the children look at their designs. Do they see identifiable figures? Suggest they use a black marker to outline these and to add features, if desired.

Variation:
Dip a twig in paint, then use it to draw on paper.

This activity also helps children learn about:
how twigs come in various shapes and sizes.

Quick Tricks for Science ©2001 Monday Morning Books, Inc.

Little Boats
A Quick Trick with Celery

Gather These Materials:
celery
vegetable brushes
plastic, serrated-blade knives
tablespoon
ranch dressing
small paper plates
blue food coloring
raisins (optional)
broccoli florets (optional)

Where: at a table

How: Have the children wash their hands thoroughly. Give each child a piece of celery. Have each child do the following: Use a vegetable brush and water to clean the celery. Cut the celery into pieces approximately as long as the child's forefinger, making celery "boats." Put one tablespoon of ranch dressing in the center of a small plate. Add one drop of blue food coloring into the dressing, and swirl it through the dressing, making blue "water." Place the boats on the water. If desired, add raisin passengers to the boats and broccoli "trees" to the shore. Eat the finished product, if desired.

Variation:
Using a cheese product in a squirt can, let each child squirt cheese into the hollow of a piece of celery and eat the finished product for a snack.

This activity also helps children learn about:
trying new foods.

Dear Parents,

In our study of plants, we're looking at plants' stems. You can continue that study at home by looking at potted plants you may have, at plants in your yard and surroundings, and at plants you pass as you walk through parks and city streets. The activity below concentrates on twigs, which are the stems of bushes and the smaller stems of trees.

A Quick Trick with Twigs

Gather These Materials:
variety of twigs and very small branches
heavy twine
ribbon

Where: outdoors to gather
anywhere to make the gift

How: With your child, gather twigs and small branches. Pick up dead wood that you see on the ground; don't break branches or twigs from bushes and trees.

Gather a bundle of twigs together in a pleasing formation and tie them snugly with the heavy twine. Wrap a ribbon around the twine and tie a bow. Give this bundle of twigs as a gift to someone who has and uses a fireplace. The twigs make good kindling.

ROOTS

This chapter provides Quick Tricks for exploring plant parts that live under the ground. The children will examine plant roots and watch them grow. They may be surprised to learn that some of their favorite foods are plant roots.

What's Under the Ground?

A Quick Trick with Weeds and Seedlings

Gather These Materials:
light blue paper
magnifying glasses
crayons/markers
field guides

Where: outdoors in an area with lots of wildflowers, weeds, and seedlings

How: Help the children pull from the ground a variety of weeds and seedlings—one for each child. Gently brush the dirt from each plant's root. Place plants on the paper to make them easier to see. Point out each plant's leaves, stem, and roots. Explain that these are parts of the plant, even the roots that grow underground.

Examine the plants' roots. Ask how they are the same and how they are different. Some plants have a branching root system, while others have a tap root—a thick root where the plant's food is stored. Let the children use magnifying glasses while looking at the roots. Look for hair-like roots coming from the main roots. Explain that all roots have these hair-like roots, which help draw moisture from the soil.

Have the children draw both kinds of root systems in their field guides. Help them label their pictures.

Variation:
Some children may pull up tree seedlings and find the seed still attached (oak seedling with acorn attached; pecan seedling with the nut attached). Discuss what happens when a seed is planted.

This activity also helps children learn about:
observing and recording observations.

Eating Roots
A Quick Trick with Carrots

Gather These Materials:
bunch of carrots with green, leafy tops attached
basin of water
scrub brush for vegetables
sharp knife, for adult use only
cutting board

Where: anywhere

How: Let your children explore fresh carrots that have the leafy tops attached. Discuss what they see. Is this a plant? How can they tell? Can they find the leaves, stems, and roots? Which part of this plant do we eat? (The root.)

Have the children scrub the carrots thoroughly. Let them watch you cut off the greens and lay them aside. Cut the carrots into carrot sticks and let everyone have a taste.

Variation:
After your children explore fresh carrots, save a few raw carrots, but wash, slice, and cook the remaining ones. Let the children taste the raw and cooked carrots. Discuss how they are the same and how they are different. Never force a child to taste or eat any food. Children are learning while they observe the carrots and hear other children's comments.

This activity also helps children learn about:
how heat (cooking) changes things.

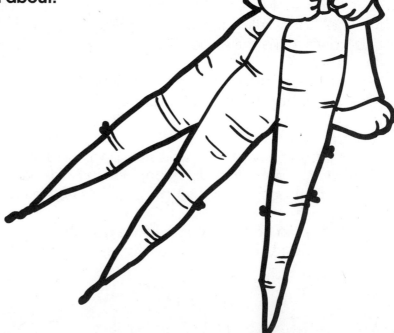

Do Roots Grow Under the Ground?
A Quick Trick with Clear, Plastic Cups

Gather These Materials:
cuttings with roots from "Growing Roots" (see page 69)
small, clear plastic, disposable drinking cups, 1 for each child
slightly larger clear plastic, disposable drinking cups, 1 for each child
potting soil
newspaper

Where: anywhere

How: Begin with the activity "Growing Roots," page 69. When the cuttings are ready to plant, have each child plant one in a small, plastic cup. Have the children draw pictures in their journals of their cuttings, including the roots. Write the day's date on the picture.
    Observe the plants for growth, observing through the plastic cups as well. Notice when the roots become visible. When they begin to fill the cup, have the children gently spill out the contents of their cups onto newspaper. Have them transplant their plants to larger cups and then continue to observe these over time for root growth. Again have them draw pictures in their journals of the plants and their roots. Date the pictures. Continue these observations and drawings as long as interest persists.

Variation:
Bring in root-bound plants to show the children. Transplant these to bigger pots or to an outdoor garden.

This activity also helps children learn about:
recording observations.

Carrot Tops and Bottoms
A Quick Trick with Raw Carrots

Gather These Materials:
fresh carrots with greens attached
knife, for adult use only
2 flowerpots
potting soil
pie tin or other shallow dish
water

Where: at a table

How: With your children, explore the fresh carrots. Identify the various parts of the plant: the leaves, stem, and root. Cut off the top of the carrot, leaving the upper 1/2 inch (1.3 cm) of the root attached.

Discuss what they think will happen if you plant the remaining bottom of the carrot in a pot of dirt. What will happen if you put the carrot top in a saucer of water?

Let the children help you plant the carrot bottom in the pot of soil. Let them put the carrot top in a shallow dish and add water. Have them observe the two experiments each day and discuss what they see. Let them add water each day to the dish and water the soil in the pot as necessary.

When the carrot top sprouts some new roots, let the children plant it in a pot of soil and continue observing it. Discuss with them what they've learned about roots.

Variation:
Let the children draw pictures to record their observations. Write their dictated words on their pictures.

This activity also helps children learn about:
a way of propagating a plant.

Quick Tricks for Science ©2001 Monday Morning Books, Inc.

Pretty Potato Vine
A Quick Trick with a Sweet Potato

Gather These Materials:
sweet potatoes, 1 for each child
strong toothpicks
wide-mouth jars
water

Where: anywhere

How: Preparation: Ask a produce manager to select sweet potatoes that haven't been treated to prevent sprouting. (Or look for sweet potatoes with a few buds or sprouts on them.)

With your children, talk about potatoes and where they grow (in the ground). The potato is a root of the potato plant, and a stem and leaves can grow from it.

For each potato, push four toothpicks into the potato, midway between the two ends. These should be equally spaced and should form spokes coming out of the potato in four directions. Place the potato in the jar with the smaller end of the potato pointing down. The toothpicks should rest on the jar's rim, suspending the potato in the jar so that it never touches the bottom of the jar.

Add water to the jar so that it covers the potato's bottom half. Set the jar and potato in a dark place for three to five days, then place the container in a sunny windowsill or in a sunny outdoor area. Have the children observe for changes. The potatoes should sprout in several places and begin growing vines.

Change the jars' water every few days to prevent bacteria growth. If a jar becomes cloudy, clean it well before returning the potato to the jar.

Variation:
When the new growth is several inches (about 6 cm) long, plant the potatoes in the ground outside. Allow several feet (1 m) of space between each to allow for spreading. Have the children watch for and measure new growth after planting the potato. After six months (or at the end of the school year) dig up the vine. What changes do you see? Do you find baby potatoes in the ground?

This activity also helps children learn about:
changes over time.

Dear Parents,

In our continuing study of plants, we have been exploring their roots. You can become part of this learning experience. While you are shopping with your child at the grocery store or vegetable market, examine the many root vegetables there. Look at parsnips, carrots, potatoes, sweet potatoes, beets, radishes, ginger, and more. Select some of these to prepare with your child and eat at home. The recipe below will give you a start.

A Quick Trick with Root Vegetables

Gather These Materials:
fresh sweet potatoes
fresh white potatoes
fresh carrots
vegetable brush
vegetable peeler
knife, for adult use only
casserole dish
onion powder to taste
salt and pepper to taste
canned vegetable juice drink

Where: in a kitchen

How: After washing hands, have your child help you scrub the vegetables with a brush and water. Point out that the dirt on the vegetables is soil the roots were growing in.

The adult peels the vegetables and cuts them in 1-inch (2.5-cm) cubes. The child puts the vegetables in the casserole dish, stirs to mix them together, and sprinkles them with onion powder, salt, and pepper. Together, pour the vegetable juice over the vegetables so it just covers them. The adult places the covered casserole in an oven at 350 degrees F (177 degrees C) for one hour. The adult removes the casserole from the oven. Serve the vegetables as part of a healthy meal.

SEEDS

Seeds are found in the "fruit" of flowering plants. First the plant produces flowers, and if conditions are right, when the petals fall from the blossom, the remaining flower parts may change into seed-bearing fruits.

Many items we call "vegetables" were once the flower part of a plant. If there are seeds inside, the item is a fruit. So squashes, pumpkins, and green peppers are actually fruits.

Gathering Seeds
A Quick Trick with Fruits and Vegetables

Gather These Materials:
a variety of in-season fruits with noticeable seeds inside
(avoid berries, bananas, kiwi,
and other fruits with tiny seeds or exterior seeds)
sharp knife, for adult use only
paper towels

Where: anywhere

How: Show the fruits to the children and help them identify each one. Help them understand that all of the fruits came from plants: apples from an apple tree, cantaloupe from a cantaloupe vine, and so on. Tell them there is something inside each fruit and challenge them to guess what is inside. Cut open the fruits to expose the seeds. Remove the seeds and let the children handle them. Discuss how the seeds are the same and how they are different. Explain that if the children plant an orange seed, an orange tree will grow; if they plant a cantaloupe seed, a cantaloupe vine will grow.

Wash the seeds and pat them dry with paper towels, then put them where the children can explore them. Cut the fruit into bite-size pieces and serve it to the children for a snack.

Variation:
Repeat the activity using tomatoes and other vegetables with seeds.

This activity also helps children learn about:
comparing and contrasting.

Gathering Seeds
A Quick Trick with Seed-Laden Plants

Gather These Materials:
paper bags, 1 for each child
large piece of white paper (bulletin board paper works well)
magnifying glasses

Where: outdoors to gather seeds
anywhere for the follow-up

How: After children discover the seeds in edible fruits and veg-etables, they will be interested in seeing seeds on plants in the ground. In early autumn when wildflowers (weeds), bushes, trees, and other plants are heavy with seeds, plan a seed-finding outing. Before you announce the outing, look for a field or wooded area with bushes, brush, weeds, and trees. A vacant lot is often home to a variety of wild plants. When you find plants in a safe location, check them to be sure that many of them have seeds. Be certain there are no plants poisonous to the touch at the site.

Gather some of the seed-laden plants and take them to class. Show them to the children and help them find the seeds. Tell the children that you will go on an outing to look for seeds like these and other seeds as well.

Take the children (and a few adults to help supervise) to the location you found. Give each child a bag. Have them explore for and collect seeds. Encourage them to find a variety of seeds instead of collecting many of one variety.

When you return to the classroom, spread a large piece of white paper on the floor. Have the children sit around the paper and display their seeds on the paper. Discuss and compare some of the finds. Use magnifying glasses for a close-up view of the seeds.

Variation:
Go to a different location on another day and gather seeds there. Examine these to see if they are the same as the seeds you gathered previously or if they are different. You will probably find different kinds of seeds by a pond, in a wooded area, and in a meadow or field.

This activity also helps children learn about:
observation.

Quick Tricks for Science ©2001 Monday Morning Books, Inc.

Seed Exploratorium
A Quick Trick with Seeds

Gather These Materials:
seeds the children have gathered
commercially available water table or a plastic dishpan
tweezers
muffin tins
egg cartons
ice cube trays
magnifying glasses

Where: at a water table (or a dishpan or tray set on a table)

How: Set up a Seed Exploratorium (a learning center for exploring seeds). Have the children bring their seeds from the "Gathering Seeds" Quick Trick (page 36) and put them in the water table (with all water removed from the table). Encourage the children to bring seeds from home to add to the collection. Also add any seeds you find while you and the children prepare snack foods: green pepper, cherry tomato, apple, pear, and so on.

Use magnifying glasses for a close-up look at seeds, and include tweezers, muffin tins, egg cartons, and ice cube trays for picking up tiny seeds and for sorting and classifying them. Share books with pictures of seeds. Encourage the children to explore the seeds, compare them, and find ones that are similar in size, shape, or color.

Variation:
Have the children help make posters by gluing various seeds onto poster board. Add the plant's name (if you know it) and a picture of the plant (drawn by the child) next to the seed. Display these near the Seed Exploratorium.

This activity also helps children learn about:
classifying.

Seed Bracelets
A Quick Trick with Masking Tape

Gather These Materials:
masking tape
large piece of white paper (bulletin board paper works well)

Where: outdoors

How: Return to the site of your initial seed hunt or go to a new location. When you get to your destination, wrap a few rounds of masking tape around each child's wrist, sticky side out. Tell the children to explore the area to find and collect individual seeds. They can put their finds on their sticky bracelets. If needed, provide a bracelet on each arm. Encourage them to find a variety of seeds. Let them compare their bracelets with their friends' bracelets. Who has the most seeds? The largest seed? The smallest? The most colors?

Variation:
Make bracelets from adhesive-backed paper, sticky side out.

This activity also helps children learn about:
where to find seeds.

Where Are the Seeds?
A Quick Trick with a Magnifying Glass

Gather These Materials:
bananas
sharp knife, for adult use only
magnifying glass
kiwi fruits

Where: at a table

How: Show the fruit to the children. Ask where the seeds might be. Cut across the fruit. Do children see any seeds? Slice the bananas crosswise and give a slice to each child. Do they see any seeds? Use magnifying glasses and look at the tiny black specks in the center of the bananas. These are the seeds.
   Repeat the activity using a kiwi. Eat the fruits for a tasty snack.

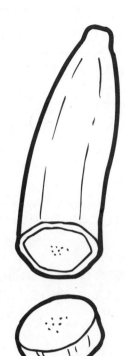

Variation:
Repeat this activity giving each child a whole, washed strawberry to explore. Give them plastic, serrated-blade knives to cut open the strawberries. Use magnifying glasses to look for seeds. Children are surprised to learn that the strawberries' seeds are on the outside of the fruit. End the activity by having the children eat the fruit.

This activity also helps children learn about:
exploring and observing to find information.

Which Has More Seeds?
A Quick Trick with Fruits

Gather These Materials:
two similar-size, familiar fruits/vegetables, such as an apple and a
green pepper; an apple and a tomato; a pear and a squash; a
zucchini and an ear of corn
sharp knife, for adult use only
spoons
marker
index cards or scrap paper

Where: at a newspaper-covered table

How: Show the children two similar-size fruits/vegetables and
have them name each one. Ask them to predict if both of these
have the same number of seeds. Ask the children to suggest
ways to find out.

Cut open the fruits or vegetables. Let the children use spoons or
their just-washed fingers to scoop out and separate the seeds. Put
the seeds from each fruit or vegetable in its own separate pile. Help
the children count each group. Use a marker and index card to
label each group and record its number.

Repeat the activity at various times throughout the year using
produce that is in season. Children never seem to tire of this activity.

Variation:
Ask children to predict if larger fruits have more seeds than smaller
ones. The adult cuts open a large apple and a small apple. Have
the children count the seeds from each. Label each group
of seeds (large apple, small apple). Repeat with other fruits.

This activity also helps children learn about:
making and testing predictions.

Quick Tricks for Science ©2001 Monday Morning Books, Inc.

Mixed Nuts
A Quick Trick with Mixed Nuts

Gather These Materials:
bag of mixed nuts
kitchen tongs
basket large enough to hold the nuts
margarine tubs—as many as the nut varieties

Where: anywhere

How: Put the mixed nuts and the tongs in the basket. Show children how to use the tongs to move nuts from the basket to the margarine tubs. Encourage them to put all nuts of the same kind together in one tub. When they finish, they'll have sorted all of the nuts.

Variation:
Put together a similar activity using tweezers and a variety of dried beans (seeds of bean plants). Have the children use the tweezers to sort the beans.

This activity also helps children learn about:
similarities and differences.

What's Inside a Seed?
A Quick Trick with a Dried Bean

Gather These Materials:
large, dried beans (lima, great northern)
clear, plastic bowl or jar with water for soaking beans
magnifying glass

Where: anywhere

How: Give each child a dried bean to explore. Discuss the way the beans feel. Tell the children the beans are seeds, and there is a surprise inside almost every seed. You will become science investigators to find out what is inside these seeds.

Gather the children's beans and put them, along with twice as many more beans, into a clear plastic bowl or jar. Cover them with water several inches (about 8 cm) above the beans, and let them stand overnight. On the following day, give each child a bean that has soaked overnight and a dry, unsoaked bean to compare. How are they the same? How are they different?

Have each child hold a bean between the thumb and forefinger. Wiggle the fingers and gently wriggle off the seed coat (the children will think it's "skin") on the outside of the bean. With the bean on its side, gently slide the sides of the bean so they separate. Inside, on one side, you'll be able to see a tiny plant embryo. Have children look at this with a magnifying glass.

Variation:
Dig up tiny seedlings and compare them with the plant embryo. Identify the tiny plant's parts.

This activity also helps children learn about:
how food for a baby plant can provide nourishment for people.

Bags of Beans
A Quick Trick with Plastic Zipper Bags

Gather These Materials:
very absorbent paper towels
pole bean seeds, several for each child
3-ounce (88 ml) cups, 1 for each child
water
plastic sandwich-size zipper bags, 1 for each child
staples
bulletin board or display board
thumbtacks
string

Where: indoors or outside for "planting"
indoor bulletin board for "observing"

How: Give each child several beans and a small cup to fill with water. Have the children place their beans in their cups to soak for a few hours.

Have each child cut or fold a paper towel to exactly fit inside a zipper bag. Have the children put the towels in their bags. While the child holds a bag steady, staple a row of staples 3 inches (7.5 cm) from the bag's top. Label the bags with the children's names.

Have the children place three beans in the stapled area of their bags so that the beans are visible between the towel and the bag's side. Add water to each bag to wet, but not soak, the paper towel. Close the bags and use thumbtacks to hang them on the bulletin board, with the children's names showing.

Watch the bags for changes; discuss and record the changes. Add water to the bags as necessary. As the beans begin to grow, unzip the bags. If desired, add strings to the board and watch the beans climb.

Variation:
After the bean plants sprout leaves and begin to climb, add plant food to the water. The plants may get blossoms and perhaps these will change into beans.

This activity also helps children learn about:
observing and recording changes.

Planting Socks
A Quick Trick with a Fuzzy Sock

Gather These Materials:
a discarded fuzzy sock for each child
plastic dishpan
potting soil or earth from your garden

Where: outdoors to collect seeds
anywhere for exploring and planting

How: Put a discarded fuzzy sock on one foot of each child, over the child's shoe. Take a walk through a wooded lot, meadow, or park. When you return, remove the socks and use a magnifying glass to look at the socks. What do you find? Where did these seeds come from? Let children explore each others' socks.

Fill the dishpan with soil. Bury one sock in the soil. Put the pan in a sunny place and water regularly. Ask the children to predict what will happen. Observe the pan daily for changes. If/when changes occur, discuss them. Continue to observe until interest lags.

Variation:
Let each child plant her sock in a class garden—a sunny area of the play yard where the soil has been turned and loosened. Label each planting with its owner's name. Water regularly; watch for changes and discuss the changes.

This activity also helps children learn about:
ways seeds travel to other locations, by catching onto people's clothing or animals' fur.

Quick Tricks for Science ©2001 Monday Morning Books, Inc.

If I Plant These, Will Birds Grow?
A Quick Trick with Bird Seed

Gather These Materials:
new kitchen sponge, 1 for each child
bucket of water
disposable dishes, 1 for each child
a mix of bird seed

Where: anywhere

How: Show the children the bird seed. What do they think it is? Ask if any of them feed birds and if they use a mix like this. Explain that birds like to eat plants' seeds, and this is a mix of seeds that birds eat. Ask the children what might happen if they plant the seeds. Discuss their answers.

Give each child a sponge. Have the children dip their sponges in the water and then squeeze out the extra water. The children put their sponges in the dishes and then sprinkle a small amount of bird seed on their sponges. Place the dishes in a sunny window and watch for changes. Discuss what happens. Also discuss where birds find food (including seed) in the wild.

Variation:
Place some seed in a bird feeder in your play yard. If you have no feeder, spill the seed into a disposable aluminum pan and place it on the ground. Watch to see which birds come to eat the seeds. See if other animals eat the seeds.

This activity also helps children learn about:
one way that plants provide food for animals.

Hairy Eggheads
A Quick Trick with Eggshells

Gather These Materials:
eggshells, one-half eggshell per child
colored permanent markers
potting soil
egg cartons
grass seed
eyedropper

Where: at a table

How: Wash empty eggshells before giving them to the children.
They can harbor salmonella germs. Give each child half of an eggshell. Supervise carefully while children use permanent markers to
draw funny faces on their shells. Have the children fill their shells
three-quarters full with potting soil. Place the shells in egg cartons
with the faces showing. One at a time, have the children sprinkle
grass seed on the soil in their eggshells. Have them cover the seeds
with a very thin layer of soil and add a small amount of water. Place
the eggheads in a sunny window. Water these tiny egg pots with
an eyedropper. Avoid over-watering. Soon the eggheads will
begin growing green hair. If the "hair" gets too long, children can
use scissors to give their eggheads a haircut.

Variation:
Have the children plant radish seeds in egghead cups for a different
kind of "hair." When the seeds sprout, remove the weakest plants
and leave one plant in each shell. When the radishes outgrow the
shells, transplant them in the garden.

This activity also helps your children learn about:
recycling materials.

Does Peanut Butter Grow on Plants?

A Quick Trick with a Blender

Gather These Materials:
roasted-in-the-shell peanuts
bowl
food processor or blender
small plastic spoons
vegetable oil (peanut oil works well)
salt

Where: near an electric outlet

How: <u>Caution:</u> Be sure to ask parents if their children have any peanut allergies. Do not do this activity in the presence of a child with peanut allergies.

The children do this: wash their hands and then shell the peanuts and place them in a bowl.

The adult does this: Place 2 cups (.5 l) of the shelled peanuts in the food processor and process them until smooth. Discuss the aroma. Show the children the processed peanuts and ask them what they think is in the processor bowl. The mixture will be thicker than commercially prepared peanut butter. Give each child a taste of this thick peanut butter. Now, add a few drops of vegetable oil and a few shakes of salt, then process again. Offer another taste to each child. Discuss the experience.

Supervise carefully. Do not let the children use the food processor.

Variation:
Place a large cutout of a blender in the center of a bulletin board. Have each child tell you about making peanut butter. Write each child's words on a sentence strip and put the strips on the board.

This activity also helps children learn about:
observing changes.

Peanut Puppets
A Quick Trick with Peanut Shells

Gather These Materials:
fresh, roasted-in-the-shell peanuts
fine-line markers

Where: anywhere

How: <u>Caution:</u> Be sure to ask parents if their children have any peanut allergies. Do not do this activity in the presence of a child with peanut allergies.

Start by having the children wash their hands. Let the children crack the peanuts and eat them, saving the shells. Have them find shell ends that will fit over their fingertips. Let them use markers to add facial features to the shells to turn them into peanut puppets. Encourage the children to let their finger puppets "talk" to each other or sing songs.

Variation:
Provide pipe cleaners, markers, glue, and other materials so the children can transform other peanut shells into animals, people, chairs, cars, and other items.

This activity also helps children learn about:
recycling nature's products.

Quick Tricks for Science ©2001 Monday Morning Books, Inc.

Dear Parents,

We are learning about plants and seeds at school. Many plants reproduce by forming seeds. When the seeds are planted, a new plant grows.

Seeds are found in the "fruit" of flowering plants. Many "vegetables" were once the flower of a plant. If there are seeds inside, technically the item is a fruit. So squashes, pumpkins, green peppers, and tomatoes are actually fruits. But no matter what we call them, all are delicious and full of healthy nutrients for us.

Why not join our exploration of seeds? You can probably find a few in your refrigerator and in your kitchen cabinets! There are seeds in apples, pears, oranges, peaches, peppers, and tomatoes. Walnuts, peanuts, and popcorn are all seeds, and so are dried beans like lentils and pinto beans. Look among your spice bottles for dill seed, peppercorns, sesame seeds, or celery seeds.

With your child, go on a seed treasure hunt in the kitchen. When you prepare fresh foods, save the seeds you find. Experiment by planting some of them to see if they'll grow. Try this "Quick Trick" with popcorn.

A Quick Trick with Popcorn

Gather These Materials:
potting soil
planting container—a clear plastic cup allows
you to watch root growth
plain popcorn kernels (not flavored, not microwave)

Where: anywhere

How: Have your child put potting soil in a planting container. With a finger, poke three holes in the soil about 1-inch (2.5 cm) deep. Gently press a popcorn kernel into each hole and smooth the soil over the top. Water, but don't over- water. Observe for changes.

A gift of flowers symbolizes the giving of love. People of all ages are enchanted by flowers' beauty. These Quick Tricks provide opportunities to explore blossoms' beauty.

Drawing Flowers
A Quick Trick with a Fresh Bouquet

Gather These Materials:
variety of fresh flowers
crayons/markers
children's field guides

Where: at a table

How: Let each child select a flower from a bouquet of fresh flowers that you've provided. (Some florists will donate flowers that are beginning to wilt.) Have the children examine their selected flowers carefully. Challenge them to draw pictures of their flowers in their field guides. Help them label the pictures. Encourage them to return their flowers to the bouquet when they are through and to select another flower to observe and draw.

Variation:
Provide artificial flowers and plastic vases so the children can make floral arrangements for the Dramatic Play Center or other areas of the classroom.

This activity also helps children learn about:
recording observations through drawing.

Pressed Posy Pictures
A Quick Trick with Fresh Blooms

Gather These Materials:
a variety of fresh blossoms (wildflowers work well)
1 paper bag for each child
box of facial tissues
discarded magazines
heavy books
white glue
water
craft stick or flat, wooden toothpicks
plain or colored paper
markers
clear, adhesive-backed paper

Where: outdoors to gather blossoms
follow-up activity at a table

How: Take your children on a blossom hunt. Walk through the neighborhood or playground, and look for blooms large and small. Give each child a bag and have the children gently pick the blossoms they find.

Press the blossoms by putting individual, flat blooms between two sheets of facial tissue, forming a "sandwich" with the flowers inside. Put the sandwiches between the pages of magazines, close the magazines, and place heavy books on top. For thick blossoms or flower heads, pluck off the individual petals, then sandwich and press them, as above.

After a week or two, remove the blossoms. Stir a mixture of equal parts of white glue and water. Use a craft stick to apply glue mixture to the dried blossoms and attach them to the paper. For tiny petals, use a toothpick to apply glue. When the glue dries, the children can use markers to add stems and vases to their pictures. Cover the pictures with clear, adhesive-backed paper to preserve them.

Variation:
Challenge the children's creativity. What are some other ways they can incorporate the dried blossoms into their pictures? Could a petal be a clown's hat? A person's nose?

This activity also helps children learn about:
how pressing between absorbent papers draws water from the flowers and dries the blossoms.

Clover Chains
A Quick Trick with Clover

Gather These Materials:
clover flowers

Where: outdoors

How: When you and your children find a patch of clover, show them how to pick the flowers at the base of the stem, close to the ground, so the flowers will have long stems.

Hold two clover flowers, A and B. Wrap the bottom of B's stem around the top of A's stem, and tie B's stem in a knot. Slide the knot as close as possible to A's flower head.

Pick up a third clover flower, C. Wrap the bottom of C's stem around the top of B's stem, just below B's flower head. Tie C's stem in a knot and slide the knot close to B's flower head.

Have the children continue in this fashion until they have clover chains long enough to wrap around their heads for crowns, their waists for belts, or their necks as necklaces. Tie the ends of the chain together to form the crowns or necklaces.

Be aware that this trick requires a high level of small muscle coordination and hand-eye coordination. Some of your children will need help. They can pick the flowers for you or other adults to tie together for them.

Variations:
Try this activity with other strong-stemmed wildflowers, such as daisies and wild aster.

This activity also helps children learn about:
small muscle coordination.

Quick Tricks for Science ©2001 Monday Morning Books, Inc.

Sunflower Hideaway
A Quick Trick with Sunflower Seeds

Gather These Materials:
garden tools
"Mammoth" or "Giant" sunflower seeds

Where: outdoors in an area that gets full sun

How: Find an area at least 6 feet (1.8 m) square in full sun. Have the children use the tools to prepare the ground for planting. With a stick, draw in the ground a circle with a 6-foot (1.8-m) diameter. Give each child a few sunflower seeds to plant somewhere on the drawn circle. Help the children space their seeds evenly. Water the seeds as needed.

After the seeds sprout and begin to grow, thin them so that the remaining sprouts are only slightly closer together than the seed package recommends. As the sunflowers grow, they will enclose the circle, forming a hideaway large enough for children to crawl into.

Variation:
Make a tepee shape by lashing together some poles at one end. Spread the other ends in a circle, and stand the tepee shape on the ground in an area that receives full sun. Weave string through the poles' tops. Plant pole beans around the outside of the tepee and train their stems up the poles and onto the strings as they grow. Long before vines cover the tepee, the children will enjoy playing in this new hideaway.

This activity also helps children learn about:
how quickly some plants grow. These large sunflowers grow very rapidly.

Miniature Vase of Flowers
A Quick Trick with an Empty Thread Spool

Gather These Materials:
commercially available, small dried flowers
scissors
white glue
empty thread spool for each child
waxed paper
6-inch (15-cm) length of ribbon for each child

Where: at a table

How: Have each child select three or four small dried flowers. Help the child cut the flower stems so they are about twice as long as the spool's height. Each flower should be a slightly different length. Stand the spool on end on a piece of waxed paper. Have the child gently gather the flowers in a bundle, with stem ends together. Dip the ends into the white glue, then put them into the spool's hole. When the glue dries, remove the spool from the waxed paper and tie a ribbon around the spool, ending with a bow. Put a dot of glue on the bow to hold it securely.

Variation:
Put a small ball of clay in the bottom of the spool and stick the flower stems into that to secure them.

This activity also helps children learn about:
the pleasure of making gifts for others.

Quick Tricks for Science ©2001 Monday Morning Books, Inc.

Dear Parents,

As part of our continuing study of plant life, we are examining and learning about flowers. There are many folk tales about flowers. Enjoy some flowers with your child as you share the folk wisdom of the "Loves Me, Loves Me Not" game below.

A Quick Trick with Flower Petals

Gather These Materials:
a flower with many distinct petals
(for example, a daisy, brown-eyed
Susan, or a wild aster)

Where: wherever you find the flower

How: Show your child how to play "Loves Me, Loves Me Not." Pick a many-petaled flower and hold it by the stem. One at a time, pick the petals from the flower, repeating as you do so, "Loves me. Loves me not." Whatever you say as the last petal is pulled from the flower is supposed to indicate your true love's feelings for you. Be sure your child understands that this is just a game that is meant for fun. A flower cannot really tell us about love.

Trees are the tallest plants. Their trunks, branches, and twigs are impressive, woody stems. Their leaves may be small and lacy or broad and thick. In addition to their beauty, they provide us with shade, fruit, and wood for our needs.

Trees Are Plants, Too
A Quick Trick with Seedlings

Gather These Materials:
small hand shovels
pastel-colored paper
magnifying glass (optional)

Where: outdoors to gather the seedlings
anywhere for examining the seedlings

How: Before the activity, locate some tree seedlings that are several inches tall. Show the growing seedlings to the children. Show the children how to remove the seedlings from the ground by digging just around the plant. Have the children dig up seedlings and gently knock dirt from the roots.

   Place each seedling on pastel-colored paper to make the plant parts easier to see. Use a magnifying glass, if desired, and explore the parts, identifying the root, stem, and leaves of each tree. Compare the seedlings. Some may have leaves, some may have needles. Compare the seedlings to grown trees.

Variation:
Replant and label a few of the seedlings/saplings. Measure their growth from time to time and see how they resemble their full-grown parent plants.

This activity also helps children learn about:
observation as a way to discover information.

Shapes of Trees
A Quick Trick with a Clipboard

Gather These Materials:
clipboards (see page 6), 1 for each child
paper
pencils, 1 for each child
commercially available field guide of trees
glue
children's field guides

Where: in a wooded area with a variety of trees

How: Take the children to a wooded area. Have each carry a clipboard with paper on it. Have the children look at the various trees and discuss the trees' shapes. Are some tall and thin? Are some triangular? Are some circular or ovoid on top? Are the trunks broad or thin?

Encourage the children to use their clipboards, paper, and pencils to draw two or three different trees. Help them find their trees in the field guide. Label their pictures with the trees' names. Help the children glue their pictures into their individual field guides. Encourage them to show and "read" their books to each other.

Variation:
Check out library books that show a variety of trees. Make these available to the children and read or show some of them at story time.

This activity also helps children learn about:
identifying trees by their shapes.

Give a Tree a Rub
A Quick Trick with Colored Chalk

Gather These Materials:
commercially available field guide to trees
paper
masking tape
crayons with the paper removed
fine-point, permanent marker

Where: outdoors

How: Before this activity, find a nearby area with a variety of trees.
Using a simple field guide, identify as many trees as you can. (You'll
need this information to label the children's rubbings.)

Take the children for a walk among the trees. Have them look at
the bark on various trees. Explain that bark protects a tree's stems in
the way that our skin and clothing protect us. Compare several
kinds of bark. Ask the children how they are similar and how they
are different.

Have each child select an interesting area of bark. Help each
tape a piece of paper over the bark. Have children hold the crayons sideways against the paper and rub up and down, pressing
firmly. Continue coloring until the paper is mostly covered.

Help children remove the tape from their pictures. Label each
picture with the marker. Compare the pictures and encourage
children to use rich vocabulary (*rough, smooth, bumpy, circular,
papery, shaggy*) to describe the patterns. If desired, write the
children's descriptive words on the papers.

Variation:
Gather the pictures and staple them into a Bark Book.

This activity also helps children learn about:
observing and naming items in nature.

How Big Around Is That Tree?
A Quick Trick with Rope

Gather These Materials:
a rope long enough to encircle the chosen trees
masking tape

Where: outdoors in a wooded area

How: Ask the children how you might use a rope to find out how big
around a tree's trunk is. Lead the discussion so that they decide to
wrap the rope around the tree at their shoulder height. Help them
do this. Use masking tape to mark the place on the rope that shows
the circumference. Show the length of rope. Now ask the children if
they think they are taller or shorter than that distance. Use the rope
to measure the children's heights then discuss the findings.
    Explore other suggestions the children had for measuring the tree.
Measure other trees in the same ways and compare their girths.

Variation:
How many children does it take to encircle a large tree's trunk if
the children hold hands and stretch out their arms while standing
close to the tree? Repeat this activity with several large trees.
See if the children can predict how many it will take to
encircle the tree.

This activity also helps children learn about:
comparing different measurements.

How Old Was That Tree?
A Quick Trick with a Tree Stump

Gather These Materials:
a tree stump that children can explore
paper
crayon

Where: outdoors at a tree stump

How: Looking at the top of the tree stump, what do the children see? (They will see concentric rings in pairs of a light ring and a dark ring.) Explain that each pair of rings represents a year of growth for a tree. Examine the rings. Are they all the same size? Explain that trees grow more in a year when weather conditions and rain amounts are favorable. Thicker rings indicate "good" years; thinner rings indicate "difficult" years. Together, count the pairs of rings. How old was the tree when it was cut down? Compare this stump and its rings with others in the area. For a permanent record of a stump's growth, lay a piece of paper over the stump's top. Hold it firmly in place while a child colors over it with the side of a peeled crayon. This produces a rubbing of the tree's rings.

Variation:
Make rubbings of stumps in other areas. Compare all of the rubbings. How were growth conditions the same? How were they different?

This activity also helps children learn about:
observing and examining items to study them.
making a permanent record of observations.

Quick Tricks for Science ©2001 Monday Morning Books, Inc.

Nature's Observers
A Quick Trick with Clipboards and Markers

Gather These Materials:
clipboards
markers or crayons
calendar
clock

Where: at a window

How: Select a tree the children can see through a classroom window. Encourage them to spend time each day or week looking at the tree. Do they see birds, squirrels, or other small animals visiting the tree? Encourage the children to use their clipboards and markers to draw pictures of their observations. Help them label the drawings, for example: "Oct. 3, 11:30 a.m., 2 squirrels chasing on the tree." "Dec. 8, 8:00 a.m., red bird in the tree." "March 22, 9:15 a.m., wind blowing the tree." (Don't expect them to be able to tell time.)

Variation:
Hang a bird feeder in the tree in a place where it is visible through the window. Encourage children to draw the birds that visit.

This activity also helps children learn about:
small changes that take place in nature.

Dear Parents,

Your child has been learning about trees: their shapes, their bark, how to measure their girth, and how to tell their age. We have been observing a tree to see if birds or small animals visit it and to see how it changes.

With your help, your child can transform a tree's small branch into a work of art. I hope that you and your child enjoy doing this activity together.

A Quick Trick with Yarn

Gather These Materials:
small branch
yarn in a variety of colors and thickness

Where: anywhere

How: With your child, look for fallen branches, knocked from a tree by the wind. Select one that forks into smaller branches and twigs and that is about the length of your forearm. Cut about 5 to 15 feet (152 to 457 cm) of each color of yarn and help your child roll each length of yarn into a small ball. Have your child do the following: Tie one end of a color onto one of the forked branches and weave this color in and out and around several of the forked branches.

Continue with each color, making a pleasing pattern. Place the branch on a table as a centerpiece or lean it against a wall as a decoration.

Quick Tricks for Science ©2001 Monday Morning Books, Inc.

NATURE'S COLORS

You can find every color of the rainbow when you look at plants. Stripes, spots, speckles, and solids—look for nature's colors in flowers, leaves, roots, stems, grasses, trees, seeds, and fruits. Mix together a fruit or vegetable salad and count the number of colors you eat.

Observing Colors
A Quick Trick with House Plants

Gather These Materials:
a variety of house plants, including some with striped, spotted, or speckled leaves and some with purple, pink, or red leaves
several blooming house plants
magnifying glasses (optional)
chart paper and a marker

Where: at a table or on the floor

How: Show the children a variety of plants. Focus on and discuss the color of the plants' leaves and stems. Compare various plants. How many colors can the children see and name? List the colors on the chart paper and ask the children to suggest a title for the chart.

Variation:
Give the children clipboards with paper. Provide crayons or markers. Encourage the children to draw close-up pictures of the leaves and blossoms they have explored, using the correct colors. If they desire, write the children's dictated words and/or the plants' names on their drawings. Gather their drawings and staple them into a "Colors of Plants" book for the classroom library.

This activity also helps children learn about:
observing to obtain information.

What Is Orange?
A Quick Trick with Paper and Crayons

Gather These Materials:
orange plant items, such as orange-hued squashes, pumpkins, tangerines, kumquats, oranges, orange flowers, fall leaves, baskets made of orange-hued straw
8.5-inch by 11-inch (21.3-cm by 27.5-cm) plain paper, 1 piece for each child
crayons, including orange
pen

Where: anywhere to explore the items and discuss the color
at a table to make the book(s)

How: Place a variety of orange plant items in your outdoor area or throughout your classroom. Talk about the color orange. Take an "orange walk," looking for items that are orange. Make a list of items the children see. Explore the items that the children can safely handle, and talk about all the items they have seen.

Fold each paper in half so it measures 8.5 inches by 5.5 inches (21.3 cm by 13.7 cm), making a four-page book. Give a book to each child. On the first page (the cover), have the children copy the book's title: "What Is Orange?" (Write this for younger children.) Have them write their names on the covers as authors and illustrators of their own books.

Provide crayons and encourage the children to draw pictures of things that are orange on the next three pages. Ask their permission to write on their pages the words they use to tell you about their pictures. (Let children who can write add their own words.)

Gather the children's books and read them to the group. Return the books and encourage the children to read them to each other and to their families.

Variation:
Repeat this activity for other colors of plant materials in your environment.

This activity also helps children learn about:
using their powers of observation.

Quick Tricks for Science ©2001 Monday Morning Books, Inc.

Colorful Quilt
A Quick Trick with Coffee Filters

Gather These Materials:
variety of flowers with stems (avoid white)
variety of individual flower petals (avoid white)
variety of living leaves
newspaper
circular coffee filters
plastic hammer
glue
construction paper

Where: outdoors to gather plant materials
at a table to transfer the colors

How: Gather live flowers, petals, and leaves from well-watered plants. Cover the table with newspaper. Place a variety of flowers, individual petals, and leaves on the newspaper. Have each child do the following: Select a number of items from the natural materials and arrange them in a small area on the table. Cover the natural materials with a coffee filter. Hammer all over the coffee filter, mashing the plant materials below so that they release their colors onto the coffee filter. Allow the filters to dry. They will look like abstract paintings.

Have the children glue their dried colored filters, bright side out, onto construction paper. When the artwork is dry, let the children help you hang everyone's creation on a bulletin board so that the display resembles a quilt. Talk about the colors in the quilt.

Variations:
Use good-quality, all-white paper towels in place of the coffee filters.
Use a flat rock or a wooden building block instead of a hammer.

This activity also helps children learn about:
hand-eye coordination.

Coloring Eggs
A Quick Trick with Onion Skins and Eggs

Gather These Materials:
eggs with white shells, 1 for each child
pot with a lid
yellow onions
water
ice

Where: in the kitchen

How: Let the children help you peel the skins from some onions (or save these from a cooking project). It doesn't matter how many you use. More skins yield a darker color. Place half of the onion skins in the pot. Let each of your children put one white egg into the pot. Discuss the eggs' color. Put the remaining skins on top of the eggs. Add water to the pot to a level 1 inch (2.5 cm) above the eggs. Let the children see and discuss the color of the water before adding it to the pot.

The adult does the following: Put the pot on the stove and bring the water to a boil. When the water boils, turn off the heat and cover the pot. Check the eggs after 20 minutes. If you want a darker color, leave them in the water longer, or keep the heat on longer. Drain the eggs, saving the cooking water for the children to see. Cool the eggs by rinsing them in cold water and then putting ice around them.

Give an egg to each child. Discuss the eggs' colors. (Some will be darker than others; some will be mottled.) Ask them where the color came from. Look at the cooking water and discuss its color.

Let the children peel and eat the eggs. Discuss their taste. Do they taste different from other hard-cooked eggs the children have eaten?

Variation:
Boil eggs in water that contains red cabbage or beets—equal amounts of vegetables and water. Predict whether the eggs or the water or both will change colors. Are your predictions correct?

This activity also helps children learn about:
colors in natural items.

Fern on an Egg
A Quick Trick with Small Plants

Gather These Materials:
fern fronds, small leaves, or small flowers
hard-cooked eggs with white shells
old pantyhose or knee-highs cut into 6-inch (15-cm) squares
scissors
twist tie
vegetable dye (food coloring, available at grocery store)
vinegar
water
small glass dishes or yogurt cups
any kind of rack—cake cooling rack, grill grid, oven rack

Where: at a table

How: Have each child do the following: Dampen a fern frond, small
leaf, or small flower and gently press it against a cooked egg's
damp shell. Holding the hosiery in one hand, lay the plant side
of the egg in the center of the square. Gently wrap the hose
around the egg, gathering and twisting the fabric in back. Use
a twist tie to secure the nylon.

Have each child do the following: Measure food coloring, vinegar, and water into a small dish following the egg-coloring directions on the food color bottle. Put the hose-encased egg in the
colored water for several minutes. Remove the egg and let it drain
on a rack before removing the hose. Each finished egg will have
a plant's imprint on its shell.

Variation:
Let the children mix various colors to create new colors. Talk about
the color changes.

This activity also helps children learn about:
how dyes work.

Dear Parents,

While studying plants, we are also focusing on their colors. You can add to the learning by observing plants' colors when you and your child are together. Compare the colors of baskets you may have at home. Compare the colors in wooden furniture you see at home or while shopping. Help your child understand that the wood in the furniture was once a part of a living tree (the trunk) and that different trees have different colors of wood. For more learning, do the "Quick Trick" below.

A Quick Trick with Colorful Plant Materials

Gather These Materials:
none

Where: anywhere

How: At any time or any place, notice a plant or an item made of plant material and say the following: "I spy with my eager eye...." Finish the sentence by describing the item, for example, "...something a lot taller than me. It's green on the top but not green all over" (a tree), "...something green and prickly" (a cactus), or "...something red on the outside and white on the inside and good to eat" (an apple). Let your child guess what you are describing. Be sure to use color names in your description.

PROPAGATION

> You and your children may have already planted seeds and watched them sprout and begin growing. These Quick Tricks show other ways of starting new plants.

Growing Roots
A Quick Trick with Scissors

Gather These Materials:
plants that propagate by rooting
scissors
plastic jars of water
later you will need: pots with potting soil, 1 for each child

Where: anywhere

How: Gather overgrown plants that propagate by rooting. (Perhaps parents or friends with plants will lend you some.) Explore the plants with the children. Tell them that these plants start new plants in an interesting way—when you cut off a piece and put it in water, the cutting will grow roots. When you plant the rooted cutting, a new plant will grow.

Cut a few inches (centimeters) of growth from vigorous shoots at overgrown places on each plant. Each cutting should have four to six leaves. Put the cuttings in labeled jars of water. Trim any leaves below the water line. Have the children watch the cuttings and describe changes they see.

When the cuttings have healthy roots, an inch (several centimeters) or longer, give each child a pot of soil. Have each child poke a hole in the pot's soil, slide a cutting into the hole, and pat soil around the cutting where it enters the soil. Again, have the children observe for changes and discuss what they see. With proper watering and light, the plants will continue to grow. Have the children take their plants home along with a copy of the Parent Letter.

Variation:
Follow the first part of this activity with "Do Roots Grow?" on page 31.

This activity also helps children learn about:
observing changes over time.

Put Me in Dirt
A Quick Trick with a Pot of Dirt

Gather These Materials:
several different plants (see description below)
pencil
pot of potting soil, 1 for each child

Where: anywhere

How: With help from a gardening center or a gardening friend, gather plants that form roots when a piece cut from the plant is placed directly in soil. Suggestions: Christmas cactus, jade plant, Swedish ivy, wandering Jew, coleus, purple heart, sedum, geranium. With the children, explore the plants.

  Let each child do the following: Break off a 3-inch (7.5-cm) long piece of one of the plants. Remove leaves from the bottom inch (2.5 cm) of the cutting. Use a pencil to poke a hole in the new pot's soil. Place the cutting into the hole, about an inch (2.5 cm) deep, and pat the soil around the cutting. Water the pot and put it in appropriate light.

  Label the pots, observe the plants over time, and discuss any changes. Have the children use a calendar to count the number of days until they see new growth on their plants. Which plants show new growth first? Plant a few extra pots of cuttings and dig them up from time to time so children can see the new roots.

Variation:
At a gardening center, ask which local outdoor or landscaping plants root in this way from cuttings. Help the children plant some of these cuttings in your school's yard. Watch for growth and changes.

This activity also helps children learn about:
the responsibility of caring for a plant.

Dividing a Plant
A Quick Trick with a Pot-Bound Plant

Gather These Materials:
a snake plant, aloe, or other pot-bound plant that can be divided
sharp knife, for adult use only
flowerpots, 1 for each child
potting soil

Where: anywhere

How: Obtain a pot-bound snake plant, aloe, or other plant that can be divided. The adult does the following: With the children watching, gently ease the plant from its pot. In some cases, you may have to break the pot to remove the plant. Place the plant on its side on the ground or on a table. Use the sharp knife to divide it from top to bottom, cutting through the plant's roots. If the resulting halves are large, cut each in half again. Repeat with this or other plants until you have a portion for each child.

Have each child do the following: Fill a pot half way with soil. Gently place a plant division into the pot, and then fill the pot with soil to 1 inch (2.5 cm) below the rim. Water and then add more soil, if necessary. Each child now has a new plant.

Variation:
Pot extra plant divisions to keep in the classroom. Watch over time. Each division will grow and spread until the new pots are full. At that point, divide again and create more new plants.

This activity also helps children learn about:
changes over time.

Bountiful Buds
A Quick Trick with Garlic

Gather These Materials:
several "heads" of garlic
pots with potting soil, 1 for each child

Where: anywhere

How: Show the children a whole head of garlic. Let them examine and explore the bumpy and papery outside of the garlic. Examine the root end and lead the children to discover that since it has roots, this must be a plant.

Show children how to break a head of garlic into individual bulbs. Let each child break off one bulb and plant that bulb in a pot of soil with the pointed end up. The entire bulb should be under the soil. The bulb will sprout long, skinny, edible leaves and perhaps, in time, show flowers. After many months, dig up the bulbs. (They will have produced more bulbs underground.) Discuss the changes.

Variations:
Plant tulips, day lilies, daffodils, or other bulbs in the school yard. Watch for their growth.
A sprouting onion from the kitchen will grow leaves and flowers if planted in a sunny spot.

This activity also helps children learn about:
observing changes over time.

Quick Tricks for Science ©2001 Monday Morning Books, Inc.

Send this note home to parents along with the plants that children have started from cuttings (see page 69).

Dear Parents,

As part of our study of plants, we have started new plants from cuttings. This is a _____ plant. Your child cut a small piece of stem and leaves from a larger plant and put the cutting in a plastic container of water. We waited for roots to emerge from the cutting, then the children planted the rooted cuttings in soil. The end result is the new plant that your child is bringing home today. You can extend your child's learning about plant propagation with this "Quick Trick":

A Quick Trick with a Plant Cutting

Gather These Materials:
the plant your child brought home today
plastic container for water (yogurt cup, juice glass)
pencil

Where: anywhere

How: Enjoy this new plant and watch it grow. After it puts out a few inches of new growth, help your child cut from the end a stem with four to six leaves on it. Place this in a plastic container of water. When roots appear and grow to about 1/2 inch (1.25 cm), have your child use a pencil to poke a hole in the potting soil beside the original plant. Gently ease the roots and stem into the hole and then pat soil in place around this new plant.

Observe the plant. Often the original plant stem will branch and put out two stems where your child cut it. With repeated cutting, rooting, replanting, and changing to a larger pot, your child can grow a large pot full of foliage.

OBSERVING CHANGES OVER TIME

Observing changes over time is an important part of scientific exploration. These Quick Tricks help your children focus on these changes. Some occur over long periods of time; some happen during a few minutes. Observing plants, noticing changes, and recording observations help your children develop thinking skills.

Changing Tree
A Quick Trick with a Tree

Gather These Materials:
camera and film
scrapbook (a loose-leaf notebook works well)
clipboards
paper
markers or crayons

Where: outdoors for observations
at a table for assembling the scrapbook

How: Have your children select a nearby tree to observe throughout the year. Visit the tree often and note any changes the children see: changing colors of leaves, falling leaves, growth buds emerging from branches, new leaves, or fallen limbs after storms, for example. Take pictures of the tree each month and put them in the scrapbook. Have the children draw pictures of the tree and its changes, using their clipboards. Place these pictures in the scrapbook, too. Record the date and time of each photo or illustration.

Variation:
Select two trees—one that sheds its leaves and one that does not. Compare them throughout the year. Discuss similarities and differences.

This activity also helps children learn about:
different ways of recording observations.

Quick Tricks for Science ©2001 Monday Morning Books, Inc.

Flowering Onions
A Quick Trick with Onions

Gather These Materials:
several onions
shovel or trowel
sunny garden area
potting soil

Where: indoors for observing onions' changes
outdoors (or in an indoor pot) for planting onions

How: As winter ends and weather begins to warm, show the children a few onions. Tell them that they will be observing the onions for a few weeks to see if they change. Keep the onions inside in a cool corner, away from sunshine and heat. Over time the onions will begin to get soft and will sprout green growth at the top. They may also show roots at the opposite end. Discuss the changes as the children discover them. Tell the children that this is a "bulb," with roots growing from its bottom and the stem and leaves growing from its top.

When the onion sprouts are 1 or 2 inches (2.5 or 5 cm) long, dig holes in the ground outdoors. Each hole should be 1 inch (2.5 cm) deeper than the onion bulb to be planted in it.

Let the children plant each onion in a hole and refill the hole with loose potting soil.

Observe the new plantings over the next few weeks. What happens? If children express surprise at the appearance of the onions' blossoms, remind them that onions are plants that grow in the ground.

Variation:
Examine tulip, daffodil, hyacinth, and other bulbs. Let your children plant these bulbs outdoors at the proper time. (Many are planted in the fall.) Watch for them to sprout in the spring. While they are growing, dig up (and then replant) one or more to examine the parts changing underground.

This activity also helps children learn about:
parts of plants.

Red Cabbage and Water
A Quick Trick with Cooking Liquid

Gather These Materials:
red cabbage
cooking pot and lid
water

Where: indoors

How: Show the children a head of red cabbage. Help them understand that cabbage is a plant, and you can eat its leaves. Let them
feel, smell, and, if desired, taste the raw cabbage. Tell them you are
going to cook the cabbage in hot water, and ask if they think any-
thing will change. Discuss the color of both the cabbage and the
water. Do they think those will change? If so, how?

Cook the cabbage, and drain the cooking liquid into a clear
bowl or beaker. Cool the cabbage by rinsing it with cold water. Put
the cabbage in a clear bowl.

Let the children see, feel, and taste (if desired) the cabbage and
the cooking water. Has the heat changed anything? (The water will
be red; the cabbage will be lighter red and it will be soft.)

Variation:
Give the children small amounts of the cooled, red-tinted water.
Encourage them to mix in other items and observe for changes.
Provide water, vinegar, blue food coloring, yellow food coloring,
baking soda, a fizzy antacid tablet. Encourage the children to
discuss their results. What happens if they mix all of their resulting
liquids? Discuss the results. <u>Caution:</u> Be sure the children don't
taste or drink any of their concoctions.

This activity also helps children learn about:
using observation skills.

Quick Tricks for Science ©2001 Monday Morning Books, Inc.

Nature's Recyclables
A Quick Trick with a Nylon Stocking

Gather These Materials:
nylon hosiery
natural materials: leaves, flowers, grass, straw, twigs
plastic materials: 6-pack holder, plastic bag, small pill bottle

Where: outdoors, or indoors using a large planting pot or plastic storage pan

How: In each nylon, place two natural and two plastic items. Bury the nylons in various places in the yard, or bury them in large flower-pots filled with potting soil. Mark the places where you have buried the items.

   If you have buried items indoors, water the pots to simulate rain, and place them outdoors in the sun whenever possible.

   At the end of three months, dig up the buried nylons. Has any-thing changed? Discuss the results. Bury the items again and dig them up once a month. Discuss what you see. Which items can nature recycle?

Variation:
Recycle plastic (and other) materials by using them as art supplies. Provide a variety of recyclable materials, along with tape and glue, and let the children use these to make new items.

This activity also helps your children learn about:
the importance of using recyclable materials.

Dear Parents,

As a part of our nature study about plants, we have been observing how plants and their products change over time. With the "Quick Trick" below, you and your child can observe the large fruit of a pumpkin plant to see how it changes after it has been cut from its vine.

A Quick Trick with a Pumpkin

Gather These Materials:
a pumpkin
newspapers
sharp knife, for adult use only
large spoon
marker

Where: pumpkin patch or grocery store to choose the pumpkin
at a newspaper-covered table to carve the pumpkin

How: With your child, select a pumpkin and decide how to decorate it. Instead of the traditional jack-o'-lantern, you might decide to carve stars, hearts, or flowers into the pumpkin.

The adult does the following: Cut the top from the pumpkin.

Help your child remove the seeds and stringy insides from the pumpkin. Use the spoon or your bare hands. Let your child draw the desired decoration on the pumpkin.

The adult does the following: Follow the child's drawing lines and cut completely through the pumpkin's flesh. Remove the cut portions.

Place the decorated pumpkin where all can admire it. Watch it over the next few days or weeks. It will begin to smell because it will start to decay. Examine and discuss the changes. Avoid breathing the odor. Before the decay becomes too bad, dispose of the pumpkin in a plastic bag or add it to your compost. Wash hands carefully after handling the pumpkin.

Quick Tricks for Science ©2001 Monday Morning Books, Inc.

MORE QUICK TRICKS

There are so many activities children can do while learning about plants. The unique Quick Tricks in this chapter don't fit into the standard categories, but they help children continue to enjoy exploring science through plants.

Recording Our Observations
A Quick Trick with Clipboards and Checklists

Gather These Materials:
variety of plants, potted or in the ground
clipboards
simple charts
clipboard-sized checklists that match the charts
pencil/marker for each child
rulers (optional)

Where: anywhere that children can observe plant life

How: Draw the children's attention to the plants, then demonstrate how to explore one plant for certain characteristics. Show the children how to mark off a simple, yes-no checklist, recording observations. Give clipboards, charts, and pencils to the children. Have each child examine a plant and then check off its characteristics on the chart. The checklist might include items like these: "has blossoms," "has more than three blossoms," "has rounded leaves," "has pointed leaves," "has more than five leaves," "has some leaves longer than my shortest finger," "plant is taller than a ruler" (or other everyday item), and "plant is taller than I am." In small groups, have the children point out their plants and share their recorded information with others.

Variation:
Have children use the clipboards, paper, and markers to draw their observed plants.

This activity also helps children learn about:
comparing their observations with others' observations.

Shucking Corn
A Quick Trick with Fresh Corn

Gather These Materials:
fresh corn on the cob, 1 ear for each child
pot of water
water for washing corn
stove for cooking, for adult use only
napkins

Where: outdoors for husking the corn
indoors for cooking
anywhere for eating

How: Have your children wash their hands thoroughly. Show them a fresh ear of corn, and let them guess what it is or what might be inside. Let them know it is part of a plant, and show them where it was cut from the corn plant. While the children watch, pull back the husks and silks of one ear of corn, leaving them attached. Let each child see the corn close up and feel the husks and the silks. Tell them that the corn kernels are the seeds of the corn plant.

   Ask the children to help you clean the remaining ears of corn. School-age children can do this with little help, but preschool children are more likely to help pull off some of the husks and silks while you hold an ear of corn and help them pull. Allow plenty of time for this task.

   When the corn is husked, wash it thoroughly and then cook it for the children to eat.

Variation:
If possible, visit a farm where the children can see live corn plants and can pick their own corn ears.

This activity also helps children learn about:
what corn looks like before it is cleaned and cooked.
what it takes to clean and prepare some fresh foods.

Quick Tricks for Science ©2001 Monday Morning Books, Inc.

Growing a Pineapple
A Quick Trick with Fresh Pineapple

Gather These Materials:
a fresh pineapple
sharp knife, for adult use only
container of sand
potting soil
water and plant food

Where: anywhere

How: Let the children explore a fresh pineapple. Have them smell and feel it. Use rich language to describe the fruit: *prickly, yellow, sticky.* Tell the children that the green things on top are leaves and the pineapple is the fruit of a pineapple plant. Ask the children if they think they can grow a new plant from the pineapple's top. Experiment to find out.

The adult cuts off the top 2 inches (5 cm) of the pineapple, including the spiked leaves. Allow the cut-off top to dry for two or three days. Have the children place it in a container of damp sand and observe it for several weeks. After a few weeks, remove it from the sand, and look at the bottom of the plant. When you and the children see roots, let the children transplant it to a pot of potting soil. Place the container where it will receive warm sunshine. Water and feed the plant. In warm weather, the plant will thrive outdoors. In time, the top may generate a new plant. The top's leaves will usually grow abundantly and roots will grow from the cut top into the soil. In time, the plant may sprout a long stem that flowers and the flowers may become tiny pineapples.

If possible, borrow a flowering or fruiting pineapple plant from a plant nursery to show the children how it grows.

Variation:
Peel the remaining fresh pineapple and serve it as a snack or as a side dish at any meal.

This activity also helps children learn about:
how pineapple looks before it is processed and canned.

Plant Parts
A Quick Trick with a Weed

Gather These Materials:
garden gloves (optional)
magnifying glass (optional)

Where: outdoors

How: Find some weeds you want to get rid of. Show the children how to pull one of the plants gently from just above the ground. This helps the roots come out, too. With the children, explore the plant(s) and identify the roots, stem, leaves, and flower (if there is one). Examine the root. Is there one thick root (a tap root), or are there many tiny roots? Use the magnifying glass for a better view. Discuss with the children why some plants are considered weeds. (A weed is a plant that is growing where we don't want it to grow. Many admired wildflowers are considered weeds when they overtake our lawns or gardens.)

Variation:
Find, pull up, and compare other varieties of weeds. Encourage the children to draw pictures of several of these in their journals. Help them label the plant parts on each picture.

This activity also helps children learn about:
an effective way of removing weeds, including the roots.

Quick Tricks for Science ©2001 Monday Morning Books, Inc.

A Gathering of Gourds
A Quick Trick with Autumn Gourds

Gather These Materials:
various gourds, 1 for each child
scissors
construction paper
glue, thumbtacks, or straight pins

Where: at a newspaper-covered table

How: Wash the dust and dirt off the gourds. Examine the gourds. Discuss ways you could turn each into a face. Should the face be happy, sad, worried, surprised, or scary? Have children cut facial features from the construction paper and attach them to the gourds to make faces. Display the faces together—a gathering of gourds.

Variation:
Substitute squashes or potatoes for the gourds.

This activity also helps children learn about:
creative thinking.

Sun Paper
A Quick Trick with Sunshine

Gather These Materials:
dark-colored construction paper, 1 sheet per child
sheets of Plexiglas as large as the construction paper

Where: outdoors on a sunny day

How: Have each child do the following: Gather leaves, twigs, and small flowers from an outdoor area. Place a sheet of construction paper in a flat, sunny area, and arrange a nature collection artfully on the paper. Place a sheet of Plexiglas over the arrangement.

Leave the papers in the sunshine for several hours—the longer, the better. The sun will bleach color from the exposed construction paper. When children remove the Plexiglas and the natural materials, they will see an image of their materials. If desired, use these papers as the covers for the children's field guides or journals.

Variation:
This activity works, too, if the papers and natural items are placed in a sunny window for a few days.

This activity also helps children learn about:
sunshine's effect on some items.

Quick Tricks for Science ©2001 Monday Morning Books, Inc.

Valentine Sachet
A Quick Trick with Potpourri Mixture

Gather These Materials:
red nylon netting, a 6-inch (15-cm) square for each child
tablespoon
dried potpourri mixture
rubber bands
narrow white ribbon, a 6-inch (15-cm) length for each child
white glue

Where: at a table

How: Put the potpourri in a plastic container, and let the children look at it and smell it. <u>Caution:</u> Be certain they avoid touching this; it could irritate the eyes if children later put their hands near their eyes. Discuss the mixture, and help children understand that it is made of dried flower petals with aromatic oils added to enhance the aroma.

Children will need help with this activity, so do this with pairs of children. Help each child do the following: Lay a piece of netting on the table. Spoon a small amount of potpourri mixture into the center of the netting, then gather the edges to the center, wrapping them around the potpourri. The adult twists a rubber band around the gathered edges. The child wraps the ribbon around to hide the rubber band and ties the ribbon in a bow. Have the child put a dot of white glue on the bow to hold it securely.

Variation:
Use different colors for the netting and ribbon to make sachets for other seasonal holidays.

This activity also helps children learn about:
recycling nature's materials.

Plant Salad
A Quick Trick with Raw Vegetables or Fruits

Gather These Materials:
raw vegetables or fruits for a salad
water and brushes for washing vegetables and fruits
plastic, serrated-blade knife for each child
kitchen knife, for adult use only

Where: in the kitchen

How: In small groups, have the children help you prepare vegetables and fruits for salads. While you work together, talk about the items you are washing, tearing, and cutting. Explore the vegetables and fruits to identify the roots, stems, leaves, and seeds. Discuss which part of each plant you will be eating: the root (carrot, beet), stems (celery, stems on fresh spinach leaves), and leaves (lettuce, cabbage). Talk about the items that are a plant's fruits and seeds (green peppers, apples, tomatoes, squash). Which parts of each will you not eat? Talk about the plants' colors and count how many colors you will be eating.

Variation:
Compare vegetables or fruits for similarities and differences. Parsnips and carrots have the same shape but different colors. Daikon radishes, horseradishes, and salad radishes are of the same family but they have different colors and sizes. Examine different varieties of lettuce; compare the sizes and colors of different varieties of apples.

This activity also helps children learn about:
the wide variety of plants that help make up a healthy diet.

Quick Tricks for Science ©2001 Monday Morning Books, Inc.

Dear Parents,

As we continue learning about plants, we'd like to share a fun idea with you. You and your child can do this activity at home, enhancing what your child has learned about plants. Let your child help you find the needed items and help choose the plants to go in them. Narrow plant choices to a few hardy growers, then let your child pick from those. This activity's finished product is sure to bring a smile to many faces.

A Quick Trick with Cast-Off Household Items

Gather These Materials:
unusual cast-off items that can serve as planters
potting soil
bedding plants

Where: anywhere

How: With your child, look around for unusual items that can serve as containers for growing plants. The basement and attic and garage sales are good places to look. Suggestions: teapot, cookie jar, cooking pot, old shoes, wagon, large toy trucks.

Together, line the bottom of the container with small pebbles and a bit of sand for drainage. Put potting soil on top of that. Plant the bedding plants in the soil and water lightly. Remember that the water has no place to drain, so don't over-water the plants.

Display the containers where others can enjoy them. Line up a convoy of flowery toy trucks or a parade of planted shoes. Share the fun!

BASIC SKILLS

These Quick Tricks provide activities that use the plant theme to teach math, language, and social skills. Many of them tell you how to make manipulative games for your children to use independently.

How Big Is It?
A Quick Trick with Ribbon

Gather These Materials:
a pumpkin
scissors
spool of ribbon
pen

Where: anywhere

How: Show the children a pumpkin. Talk about its circumference: the measurement of how round it is at its largest point. Tell them that they will each estimate how much ribbon it will take to exactly go around the pumpkin's circumference.

Have each child cut of a length of ribbon that he thinks will exactly go around the pumpkin. Write the children's names on their ribbons. Have the children, in turn, wrap their ribbons around the pumpkin. Whose was too long? Whose was too short? Whose was just right?

Tape the ribbons to a chart to make a display of "Too Long," "Too Short," and "Just Right." If no one's ribbon is "just right," cut a ribbon the exact length of the circumference to mount on the chart. Arrange the ribbons from shortest to longest with the "Just Right" ribbon between the two groups.

Variation:
Repeat this activity often using other vegetables or fruits such as watermelon, grapefruit, and apples. Repetition improves estimation skills. Use yarn or string instead of ribbon.

This activity also helps children learn about:
comparing.

Pumpkin Patch
A Quick Trick with a File Folder

Gather These Materials:
file folder
permanent markers
scissors, for adult use only
orange construction paper
pictures of items that begin with consonant sounds you are
practicing in this game
clear, adhesive-backed paper
glue
manila envelope

Where: at a table for the adult to make the game
anywhere to play the game

How: Open a file folder and draw on it five green vines with leaves.
These are pumpkin vines. From construction paper, cut out pumpkin
shapes in appropriate sizes to go with the vines, about 1 1/2 to 2
inches (3.8 to 5 cm) in diameter. Make 15 pumpkins, 3 for each
vine. On each vine, print a consonant letter. (Use letters that don't
look similar: B, S, T, W, and D would be good; B, P, and R are too
similar in shape as are M, W, and N. On each pumpkin, draw or glue
on a small picture starting with a consonant sound that matches a
vine's label. Try *sun, sink,* and *scissors* for the three S pumpkins and
tree, table, and *taco* for the three T pumpkins. Use clear, adhesive-
backed paper to laminate the pumpkin shapes.

Put the pumpkin shapes in the envelope, and glue the envelope
to the outside back of the file folder. The envelope's flap should be
open and should face the outside. Draw some pumpkins and vines
on the front of the folder.

To use this game, a child takes the pumpkin shapes from the
envelope, opens the folder, and lays it on a flat surface. The child
matches initial sounds to the appropriate consonants to put the
pumpkins on their correct vines.

Variation:
Make a similar file folder game writing a numeral on each vine and
having the player put the correct number of pumpkin shapes on
each vine.

This activity also helps children learn about:
working independently.

Rhyming Salads
A Quick Trick with Vegetable Shapes

Gather These Materials:
adult scissors, for adult use only
markers
plain paper
4 margarine tubs
glue
construction paper
clear, adhesive-backed paper
manila envelope
small basket

Where: at a table for adult to make the game
anywhere to play the game

How: Cut four circles of plain paper slightly smaller than the bottom of the margarine tubs. On each circle, draw a picture of one word from each word family from the following list of rhyming word families. Glue one circle in the bottom of each tub.

Cut a variety of vegetable shapes from construction paper (bell pepper, carrot, potato, radish, turnip, celery). On each vegetable, draw one picture from the remaining rhyming words, being sure that each margarine tub has one or more rhyming matches. Label the envelope "Rhyming Salads."

Laminate the envelope and the vegetable shapes with clear, adhesive-backed plastic. Place the vegetable shapes in the envelope and the envelope and margarine tubs in the small basket.

To play, a child spreads out the tubs (salad bowls). The child looks at the picture on a vegetable shape and then puts that vegetable into the salad bowl with a rhyming picture. Remind children to "mix up" the vegetables when they put the game away so the game will be challenging for the next player.

fox	hat	fan	log
socks	cat	man	dog
blocks	bat	pan	frog

Variation:
Use commercially available rhyming pictures (available at teacher supply stores) and mount them on the vegetable shapes.

This activity also helps children learn about:
working independently.

Letter Leaves
A Quick Trick with Leaf Shapes

Gather These Materials:
adult scissors, for adult use only
green construction paper
permanent markers
manila envelope
clear, adhesive-backed paper

Where: at a table to make the game
anywhere to play the game

How: The adult does this: Cut eight different pairs of matching leaf shapes from the construction paper. Put leaves in pairs. On one leaf of each pair, write an uppercase letter; on the matching leaf, write the matching lowercase letter. Continue with the remaining leaves, using different letters for each pair. On the envelope, draw two matching pairs of leaves. Write matching uppercase and lowercase letters on each pair. Laminate the leaf shapes and the envelope with clear, adhesive-backed paper for durability.

Have the child take the leaves from the envelope, mix them up, then spread them out. Have him look at the leaves and then put the pairs together. Before putting the game away, have him ask another child to check his work. (This gives another child an opportunity to practice the matching skills, also.) Have the child mix up the leaves before putting away the game.

Variation:
For younger children, use all uppercase letters. The child matches two leaves that both have the same uppercase letter written on it.

This activity also helps children learn about:
working independently.

How Does Your Garden Grow?
A Quick Trick with Buttons

Gather These Materials:
9-inch by 12-inch (22.5-cm by 30-cm) poster board
markers
variety of buttons
small plastic zipper bag
large plastic zipper bag

Where: at a table to make the activity
anywhere to play the game

How: Gather a dozen buttons in a variety of shapes and sizes. Turn the poster board so that it is 9 inches (22.5 cm) tall and 12 inches (30 cm) wide. Use a brown marker to draw "garden soil" on the bottom 2 inches (5 cm) of the poster board. Draw 12 stems with simple leaves growing out of the soil. Place a button at the top of each stem, and trace around all of the buttons. (These are the flowers.) Place the buttons in a small plastic zipper bag and place both the poster and button bag in a larger plastic zipper bag. To play the game, a child puts the buttons in the correct tracings.

Variation:
To increase the difficulty, make several game boards and store them and their buttons all together. This game can be played by one or more children.

This activity also helps children learn about:
paying attention to detail.

Quick Tricks for Science ©2001 Monday Morning Books, Inc.

Flower Vases
A Quick Trick with Flowering Weeds

Gather These Materials:
10 plastic bottles to use as vases (discarded dishwashing liquid bottles work well)
permanent marker

Where: outdoors to gather the weeds or flowers
anywhere to fill the vases

How: Gather 10 plastic bottles to use as vases. With the marker, write the numerals 0 to 9 on the vases, writing one numeral per vase. Encourage the children to gather flowering weeds or wildflowers. Put all the flowers and the vases together, then let the children, in turn or in small groups, put the correct number of flowers in each vase.

Variation:
Substitute fabric flowers for the fresh ones and make this activity a permanent part of your Math Center.

This activity also helps children learn about:
recycling plastic bottles.

Watermelon Slices
A Quick Trick with Construction Paper

Gather These Materials:
green, pink, and white construction paper
adult scissors, for adult use only
glue
black marker
clear, adhesive-backed paper
plastic zipper bag

Where: at a table to make the activity
anywhere to play the game

How: The adult does this: Cut five 6-inch (15-cm) circles from the green paper and five 5-inch (12.5 cm) circles from the pink paper. Cut the circles in half, yielding 10 half circles of each color. Glue one pink circle onto each green circle, lining up their straight edges, so that each resulting item looks like a half-slice of watermelon. Draw one black seed on one slice, two black seeds on another slice, three on another, and so on up to nine seeds. Leave one slice with no seeds.

Cut 10 four-inch (10 cm) squares from the white paper. Write a numeral, 0 to 9, on each square, writing one number per square. Cover the slices and the number cards with clear, adhesive-backed paper for durability. Place all of the materials in a plastic zipper bag and show the children how to play the game.

The player spreads out the watermelon slices in front of himself. He chooses a white number card, reads the numeral, then places the card beside the slice that has that number of seeds. The player continues until all of the numeral cards have been placed beside their matching slices.

Variation:
Show the children how to place the number cards in order from 0 to 9, and then have them place the matching slice beside its number card.

This activity also helps children learn about:
reading and understanding numerals.

Flowers Come and Flowers Go
A Quick Trick with Fabric Flowers

Gather These Materials:
scissors, for adult use only
fabric flowers (inexpensive at garage sales)
12-inch (30-cm) poster board square for each child

Where: at a table to make the activity
anywhere to play the game

How: Cut the fabric blossoms from their stems and discard the stems. Give each child a poster board square to use as a game board and a handful of fabric flowers. Tell simple, two-sentence stories to the group and have them put flowers onto their boards or remove them in response to each story.
    Example:
        "Three flowers grew in Tonya's garden." (Children put three flowers on their game boards.) "She picked two for her grand mother." (Children remove two flowers.) Have the children clear their boards for the next story.

    Resist the temptation to have the children "solve" the problems and tell you how many flowers remain at the end of each story. The learning comes from acting out the problem. After many experiences with the game, the children will begin telling you the "how many" part that could come next in the story. Follow their cues, and avoid rushing into this part.

Variation:
Remove the jokers, aces, and face cards from a deck of cards. Put the remaining cards with the flowers and game boards from the game. Suggest that children take turns turning over a card and telling the first line of a story using that number of flowers. Then the child makes up the second sentence of the story without a card prop. All of the children add or take away flowers based on the stories.

This activity also helps children learn about:
listening for details.

Sorting Leaves
A Quick Trick with Leaf Rubbings

Gather These Materials:
a leaf from each of 8 different kinds of trees
plain paper cut in 8-inch (20-cm) squares
crayon with the paper peeled off, or a colored pencil
tagboard (or old file folders) cut into sixteen 8-inch (20-cm) squares
glue
clear, adhesive-backed plastic

Where: at a table to make the game
anywhere to play the game

How: The adult does this: Place a leaf on the table and place a square of plain paper on top of it. Color back and forth over the paper with the side of the peeled crayon. An image of the leaf will appear. Repeat with this leaf and another square of paper. Make two rubbings of each of the remaining leaves. Glue each rubbing to a square of tagboard. For durability, cover each rubbing with clear, adhesive backed plastic.

Show the children how to play the game. Have a child spread out the cards and then find the matching cards and put them together. Have him ask a friend to check his work.

Variation:
Gather leaves of three sizes from three or four different species of trees. Make a rubbing of each leaf, and make a game as above. The player matches rubbings of leaves from the same tree.

This activity also helps children learn about:
noticing visual differences and similarities.